Exploring Science in Ancient Indian Texts

Exploring Science in Ancient Indian Texts

Edited by

Bal Ram Singh

Center for Indic Studies
Dartmouth (USA)

DK PRINTWORLD
Publishers of Indian Traditions

Cataloging in Publication Data — DK
[Courtesy: D.K. Agencies (P) Ltd. <docinfo@dkagencies.com>]

Exploring science in ancient Indian texts /
edited by Bal Ram Singh.
p. cm.
Includes bibliographical references.
ISBN 9788124607527

1. Science – India – History. 2. Science in literature.
3. Vedas and science. 4. Vedic literature – History and criticism. I. Singh, Bal Ram.

DDC 294.5921015 23

First published in India, 2014
ISBN 13: 978-81-246-0752-7

Published by:
Center for Indic Studies
University of Massachusetts, Dartmouth
285, Old Westport Road
Dartmouth, MA - 02747 (USA)
Phone: 508-999 8588 *Fax*: 508 999-8451
Email: bsingh@umassd.edu *Website*: www.umassd.edu/indic

and

D.K. Printworld (P) Ltd.
Regd. Office: Vedaśrī, F-395, Sudarshan Park
(Metro Station: Ramesh Nagar)
New Delhi – 110 015
Phones: (011) 2545 3975; 2546 6019 *Fax*: (011) 2546 5926
E-mail: indology@dkprintworld.com
Website: www.dkprintworld.com

Printed by: D.K. Printworld (P) Ltd., New Delhi

Preface

EXPLORING science to understand the mystery of the world we live in is a curiosity man has pursued since time immemorial. It has not necessarily been called science, rather it is many times known as philosophy, culture, traditions, etc., but something that has been based on rational analysis of objects or events. Consequently, an understanding of the mystery of this world is not spoken in a uniform manner. It is not just due to the lack of a common language, but rather due to the lack of a common experience. The narrative of modern science although founded on many of the knowledge streams from entire world has somehow has been narrowed, at least in perception, to the Western world crafted mainly by the European societies. India, arguably the only living ancient civilization in the world, has a grand narrative of not only science and technology but also ontology and the epistemology of science, which is more sophisticated and substantive than what we pursue in even in the most fundamental aspects of modern science. Therefore, it is essential to bring such a narrative to expand the horizons of science, for the benefit of society as well as the science itself.

Since becoming the Director of the Center for Indic Studies at the University of Massachusetts Dartmouth in 2000, I have had many occasions to listen to lectures, interact with scholars, and organize conferences on topics closely related to India's traditions and their connection to science and technology, both ancient and modern. In most cases, while the information was attractive it always left me feeling scanty on the robustness of the argument. There never appeared to be deliberations on multiple aspects of applying modern yardsticks of science to examine the claims and references to science and technology in ancient Indian texts. There was never adequate opportunity to debate it to understand and advance the concepts. At that point I availed an opportunity under a funding from BVM Foundation to organize a workshop-cum-retreat on Exploring Vedic Sciences in November 2009 for roughly four days, inviting scholars who have developed substantial concepts based on theories and experiments, and providing them about 4 hours each to present their topics in a classroom format with adequate time for discussion and debate. The scholars were asked to submit their concepts in the form of chapters in a textbook format so that anyone reading them can follow the material for learning, not just for reference.

The chapters were reviewed by experts in the field, and subjected to revision, before final acceptance for publication. It is hoped that the detailed ideas presented in this book will trigger further research and scholarship not only on the topics covered but also will inspire discussions on many other topics not only from the traditions of India but also other ancient traditions of the world. There is an urgent need of expansion of concepts based on experiences of people in different parts of the world, multiplying the intellectual resources, and then integrating the knowledge to broaden the scientific foundation.

I am very grateful to Mr Mukesh Chatter and Mrs Preeti Chatter of BVM Foundation for inspiring us to embark upon this project, by providing the funding. I am also thankful to my staff, particularly Ms Maureen Jennings, for making all the arrangements for the workshop, and my students who were participants not only as a group of small audience but also providing musical entertain=ment to workshop participants.

Finally, I am indebted to Mr Susheel Mittal of D.K. Printworld, for taking keen interest in the publication of this book.

Bal Ram Singh

1. Speakers and participants of the Workshop

2. Workshop sessions

3-5. Workshop sessions

6. Professor MG Prasad demonstrating acoustics of conch

7. Professor Prasad, Ms. Koyel Ghosal (participant), Professor Singh, Dr. Nagarajan Thirunavukkarasu (participant)

8. Ms. Ghosal demonstrating the art and science of Indian music

9. Ms. Ghosal playing sitar for Workshop participants

Introduction

Bal Ram Singh

Science and Philosophy

SCIENCE in modern world is defined in simple terms as a systematic study of a topic, subject, or a problem that provides consistent results. Newtonian definition of scientific truth requires it to be non-falsifiable, this introducing a quality control for systematic study of the topics. In requiring the scientific truth to be non-falsifiable it is inherent that the pursuer of the science makes attempts to falsify the emerging truth, and accept the questions and falsification efforts by others, and only when it is possible to falsify the facts and conclusions as scientific until proven false in the future.

There is a great deal of concern in the scientific process and purpose. Since the process can be influenced based on the individual capabilities broadly defined as mental and physical tools employed in the process, the scientific research is suggested to focus on the object and thus be objective, rather than subjective. The purpose of science could vary from genuine curiosity to commercial interest, opening it to subjective bias. Thus objectivity of information may become compromised. The scientific pursuit is therefore ideally carried out in an unbiased manner. The question, however, is that whether it is possible to employ unbiased approach to observations.

Prof. F.S.C. Northrop in his foreword to Werner

Heisenberg's book, *Physics and Philosophy* (Heisenberg, 1999), wrote:

> The instruments of modern science derive from its theory and require a comprehension of that theory for their correct use.
>
> This theory rests on philosophical as well as physical, assumptions.
>
> When comprehended, these philosophical assumptions generate a personal and social mentality and behaviour quite different from, at points incompatible with, the family, caste and tribally centred mentality of native Asian, Middle Eastern or African people.

Clearly, philosophical and cultural influence could play a crucial role on scientific theories and data interpretations. Therefore, both cultural and philosophical background of scientific ideas is an important factor to consider when assessing science and technology. There is a clear distinction in the culture and philosophy of the Eastern and Western world, in large part due to the geographical and climatic reasons. For example, during ice age, there is difficult life in the northern hemisphere, which makes survival a main goal of life. This climatic reality triggers a series of cultural traditions which influence derivative philosophies of life. The Western culture is more impacted by this phenomenon than the Eastern culture, particularly of the Indian subcontinent, as the latter is surrounded by the Himalayan range of mountains on one side while by the ocean on the other. Indian cultural traditions are thus more likely to continue, and leading to more sustaining philosophies.

Vedic knowledge and its acclaimed eternity result perhaps this ontology, and its philosophical traditions have had extended period of time to be tested and tried. Are these philosophies amenable to modern practices of science? Vaiśeṣika philosophy principles are quite compatible with

Contents

Preface v

Introduction 1
— Bal Ram Singh

1. The Case for Non-dual Science Bringing the Mystery back 12
— Jerry Solfvin

2. Indian Tradition, Science, and the Problem of Consciousness 42
— Subhash Kak

3. Vedānta and Neuroscience 68
— Vinod D. Deshmukh

4. The Mystery of Time Converging Thought Currents from Vedānta, Physics, and Neurosciences 130
— Raju Chidambaram and Shobha Chidambaram

5. Consciousness Normalization 173
— Ravi Jata

6. A Proposed Model to Quantitatively Assess Consciousness: Employing Time and Temperature Relationship to Mind 231
— Bal Ram Singh, Rishi Bal Singh and Sukalyan Sengupta

7. Vedic Perspectives on Acoustics — **M.G. Prasad** 257

8. Siddhānta-Darpaṇa of Nīlakaṇṭha Somayājī — **K. Mahesh and K. Ramasubramanian** 282

List of Contributors 317

Index 321

modern chemistry and vice versa (Singh 2003). *Vaiśeṣika-Sūtras* begins with the definition or purpose of the *dharma – yato 'bhyudaya-niḥśreyasasiddhiḥ sa dharmaḥ*, meaning "*dharma* is the one which lifts up everything to the greatest realization (salvation or *mokṣa*)". In other words, anything that is destined to life (existence) and death has to follow the *dharma*. *Padārtha* is created by the *dharma*, and subsequently *padārtha* expands itself into different entities and categories encompassing the universe we know as made of the substance or matter.

Modern science as we know today although has had many of its roots in Vedic and Upaniṣadic knowledge, its current concepts were deliberately dissociated even from philosophical bases, mostly due to the historical reasons. After European Renaissance during 1400s to early CE 1500, there was a major introspection in the Western world largely under the dark shadows of Church, which controlled the Western society in those times (Kirkpatrick 2002). While in India where Vedic practices, even if these were not perfect or genuine, still dominated the society, it was under foreign rule with very limited freedom to deal with such issues. In the European Renaissance, looking back to Greek philosophies and even Roman Empire provided inspiration. Consequently, scientific theories and interpretations had a clear bias with respect to philosophy and culture.

It is important to note that modern scientific methods and fields of learning to understand ancient knowledge may not be compatible, especially when there may be major cultural and philosophical differences between the society where these scientific tools are developed and the society of the ancient knowledge. Therefore, examination of the basic premises in the Western philosophy or science is needed to compare them with some of the fundamental assumptions in the Indian philosophical systems.

Western thoughts (Agrawal 2001) hold that —

1. man cannot know metaphysical truths by direct experience,
2. even if they can be known, so far no man has known them, and
3. being pure speculations, various schools of Indian thoughts, like the speculative system of the West, must be mutually contradictory, and that if one is true then all the others must be false.

Presuppositions for Indian thoughts are:

1. Man can know metaphysical truths directly,
2. there have been such men and there may still be seers with such knowledge,
3. seers (*ṛṣis*) teach metaphysical truths after knowing them directly, and
4. while all the *ṛṣis* know the same truths, they teach these truths in different standards represented by the texts of the schools (Agrawal 2001).

In modern science scientific truth is something that cannot be falsified. There is no need for direct evidence to prove it true. Elements of modern science are fortified by the requirement of objectivity in pursuit of scientific enquiry and experiments.

On the surface it appears that there could be a major contradiction between the Vedic knowledge and the Westernized modern science. However, in practice, at least at the fundamental level, science follows the Vedic ways of pursuing the truth. For example, scientific pursuit embraces the idea of knowing the truth in more than one ways, and actually scientists find more than one way to prove their points or ideas, be it more than one technique, organisms, or even researchers. Therefore, in the current time of scientific predominance of learning, it is incumbent upon scholars to use the science itself to examine some of the intricacies of the ancient knowledge.

Influence of Vedic System on Modern Science

The influence of Vedic thoughts on modern science may not be considered obvious for a variety of reasons. One, during and after the European Renaissance, people of scientific temper decided not to confront Church authorities and have mostly continued to keep their own religious devotion and scientific analysis of the nature separate. So, there was no effort to associate the knowledge with any religion, which is how Western world looks at the Vedic ideas. Two, Western political as well as religious authorities embraced scientific knowledge for their own purposes to remain a dominating force. In fact, since 1936 Vatican conducts a biannual conference on science and religion to promote the progress of mathematical, physical and natural sciences, and to study epistemological problems relating thereto. Three, people historically related to the Vedic culture have remained subjugated at least mentally if not physically to assert their intellect in recognizing them. Four, Western scientists and historians themselves consciously or unconsciously have suppressed some of the obvious references to the Vedic thoughts in the development of modern sciences.

However, some of the great luminaries of the twentieth-century science have clearly stated the Vedic thoughts to exactly explain the highest ideas of the modern science. Erwin Schrödinger, one of the major pillars of modern quantum mechanics, writes:

> Within a cultural milieu (*kulturkries*) where certain perceptions (which once had or still have a wider meaning among other peoples) have been limited and specialized, it is daring to give to (my conclusion arrived at by putting

> together quantum mechanics with biology) the simple wording that it requires. In Christian terminology to say: "Hence I am God Almighty" sounds both blasphemous and lunatic. But please disregard these connotations for the moment and consider whether the above inference is not the closest a biologist can get to proving God and immortality at one stroke.
>
> In itself, the insight is not new. The earliest records, to my knowledge, date back some 2,500 years or more. From the early great Upaniṣads the recognition *ĀTMAN = BRAHMAN* (the personal self equals the omnipresent, all comprehending eternal self) was in Indian thought considered, far from being blasphemous, to represent quintessence of deepest insight into the happenings of world. The striving of all the scholars of Vedānta was, after having learnt to pronounce with their lips really to assimilate in their minds, this grandest of all thoughts.
>
> Again, the mystics of many centuries, independently, yet in perfect harmony with each other (somewhat like particles in an ideal gas) have described, each of them, the unique experience of his or her life in terms that can be condensed in the phrase: *DESU FACTUS SUM* (I have become God) . . .
>
> — Schrodinger 1964

In Einstein's view, experiencing the notions of space and time lie at the foundations of his theory of relativity (Ananthu 1992). According to Einstein, "For a physicst, the distinction between past, present, and future is an illusion, though a stubborn one". And this was not anything less than a true science (Ananthu 1992). Einstein had a very strong opinion on the subject of mystic, science and religiousness:

> The most beautiful and most profound emotion we can experience is the sensation of the mystical. It is the sower of all true science. He to whom this emotion is stranger, who can no longer wonder and stand rapt in awe, is as good as dead. To know what is impenetrable to us really exists,

> manifesting itself as the highest wisdom and the most radiant beauty which our dull faculties can comprehend only in their primitive forms — this knowledge, this feeling is at the centre of true religiousness.
>
> — Ananthu 1992

At the same time there have been prominent scientists who totally believe in reductionist theory of view to understand and explain the nature of living and non-living, and show even modest contempt towards religious way of realizing the truth. Nobel laureate Francis Crick who, on one hand, was so much affected by Erwin Schrödinger's famous book, *What Is Life,* that he changed his field from physics to biology (Satya and Umesh 1992), he still took a very different point of view of understanding life, primarily by reducing everything to the smallest molecules. Crick (1993) observes:

> Most of the religious beliefs we have today originated in a time when the earth, while a small place by our standards, was then thought of a being very large, even though its exact extent was unknown. Anyone person had direct knowledge of only a tiny part of it. It was not implausible to believe that this large earth was the centre of the universe and that man occupied the leading place in it. The earth's origins seemed lost in the mists of time and yet the span of time thought to be involved, while it seemed long in terms of human experience, we now know to be ridiculously short. It was not implausible to believe the earth was less than 10,000 years old. We now know its true age is about 4.6 billion years. The stars seemed far away, fixed perhaps in the spherical firmament, but that the universe extended as far as it does — more than 10 billion light years — was almost inconceivable. (An exception has to be made here for certain Eastern religions, such as Hinduism, that take pleasure in inflating times and distances for the sheer joy of it.)

It is obvious that there are scientists like Crick who are very hostile to religious groups of all sorts without willing to examine the ancient knowledge itself. However, even he had to acknowledge the contribution of Vedic knowledge of the concept of time, though reluctantly.

Rationality and Practicality of Vedic Traditions for Modern Science

With the realization of the limitation on human sensual faculties, scientists have adopted models to examine and understand the matter as a substance. These models have been developed, and are acceptable (Barton 1997) because they

a. describe the new in terms of the old,
b. express unfamiliar in terms of the familiar,
c. relate the abstract to the concrete, or
d. integrate several previous models in an intuitively satisfying way.

It is notable that with the application of these models we acknowledge as scientists that we are not even dealing with reality, just a simple perception of it. At best, these objects in the solid world around us are really only images or approximations around us (Stapp 1982). Our scientific theories are just theories, their relationship to reality again depends on the perception. Even the instrumental measurements are based on a perceived reality. Echoing these sentiments, Stephen Hawking said in a 1994 lecture: "I don't demand that a theory correspond to reality because I don't know what it is" (Hawking and Penrose 1996).

Given the above limitations, it is still possible to model reality in terms of the modern scientific languages. For understanding the *dharma* paradigm, I believe the scientific language is the most appropriate means of communication for the world at large. Whereas science may not have solved all the problems in the world, the scientific approach has become the universally accepted way of learning and understanding. Moreover, the tradition of *dharma* is very amenable to scientific inquiry and examination, and its rational and logical features openly embrace scientific queries, without being offended (Singh, 2011). At the same time, one does not have to verify every idea or practice in *dharma* or other traditions with science to consider them valid. I only suggest that the language of science and scientific approach be used in the debates and discussions about *dharma* or Vedic traditions.

In fact, it is quite possible and important to use many of the concepts of *dharma* traditions to either rationalize or even reveal new concepts on some of the fundamental questions being addressed in science. These include origin of matter and universe, origin of life, origin of species and evolution, state of consciousness and mind, etc. The chapters presented in this book address three types of topics — the basic premise of scientific approach to examine reality (Solfvin — The Case for Non-dual Science; Kak — Indian Tradition, Science, and the Problem of Consciousness; Jata — Consciousness Normalization), mathematical and scientific knowledge, derivation, and application to Vedic texts (Mahesh and Ramasubramanian — Siddhānta-Darpaṇa of Nīlakaṇṭha Somayājī, Jata — Consciousness Normalization), and models for current scientific issues with Vedic perspective (Prasad — Vedic Perspectives on Acoustics; Chidambaram and Chidambaram — The Mystery of Time; Singh, Singh and Sengupta — A Proposed Model to Quantitatively Assessing Consciousness; Deshmukh — Vedānta and Neuroscience). These topics cover ideas of matter and universe, consciousness and mind, and fundamental questions of defining and applying science and scientific approach.

There are more attractive aspects of Vedic knowledge which are becoming practical for human health. These include Āyurveda and *yoga*. Although Āyurveda and *yoga* are being examined with herbal chemistry and neuroscience, these approaches remain limited in the absence of fundamental assumptions of Vedic concepts. The efforts made by authors in this book are to lay down some of the fundamental principles of Vedic traditions and practices. Since ancient Indian texts have been shown to contain many advanced technologies and scientific advancements (Singh, Jha, Singh and Mishra 2011), one wonders what scientific methodologies have been in practice in India, and whether any of those methodologies are relevant in modern sciences. Vedic philosophy expects the knowledge seekers and practitioners to remain detached and selfless to make observations. *Pātañjala Yoga-Sūtras* deliberates extensively on how to approach knowledge by being totally detached, and *yoga* practices of *dhyāna* (meditation) develops natural ability to observe events and extract knowledge in an unbiased way, something that is certainly the professed goal of modern science. It is therefore hoped that this book will lay the initial foundation for improving modern science through practising and integrating Vedic traditions, as outlined in many ancient Indian texts.

References

Agrawal, M.M., 2001, *Six Systems of Indian Philosophy*, Delhi: Chaukhamba Sanskrit Pratishthan.

Ananthu, T.S., 1992, "Does God Play Dice?", in Narlikar et al. (eds.) *Philosophy of Science*, Delhi: Munshiram Manoharlal, pp. 39-48.

Barton, A., 1997, Chapter 1, *States of Mind States of Matter*, Bristol: Institute of Physics Publishing.

Crick, F., 1993, Chapter 1, *The Astonishing Hypothesis: The Scientific Search for the Soul*, New York: Charles Scribner's Sons.

Hawking, S.W. and R. Penrose, 1996, "The Nature of Space and Time", *Scientific American*, **275**: 44-49.

Heisenberg, W., 1999, *Physics and Philosophy: The Revolution in Modern Science*, Amherst, N.Y.: Prometheus Books.

Kirkpatrick, R., 2002, European Renaissance 1400-1600, New York: Longman.

Satya, Y. S. and G. Umesh, 1992, "Chance and Determinism in Physics", in

Narlikar et al. (eds.) *Philosophy of Science,* Delhi: Munshiram Manoharlal, pp. 96-117.

Schrodinger, E., 1964, *What Is Life?*, Cambridge: Cambridge University Press.

Singh, B.R., 2003, "Use of Chemistry to Understand Vedic Knowledge", in B.D. Sharma (ed.) *Contemporary Views on Indian Civilization,* Atlanta, CA: World Association of Vedic Studies, pp. 388-99.

———, 2011, "Scientifying the Science: The Art of Making Everything a Science", in B.R. Singh, G. Jha, U. Singh and D. Mishra (eds.) *Science and Technology in Ancient Indian Texts,* New Delhi: D.K. Printworld, pp. 1-10.

Singh, B.R., G. Jha, U. Singh, and D. Mishra (eds.), 2011, *Science and Technology in Ancient Indian Texts,* New Delhi: D.K. Printworld.

Stapp, H.P., 1982, "Mind Matter and Quantum Mechanics", *Foundations of Physics,* **2**: 363-99.

1

The Case for Non-dual Science
Bringing the Mystery Back

Jerry Solfvin

Abstract

Dualism — the breaking of ideas, concepts, and theories into "twos" — is a tool that is useful in science. Like the many logical riddles built upon the **city of truth-tellers** and the **city of liars**, which is good logical exercise but has little direct applicability in real-life settings where there are no such simplistic, extreme, polar communities, there is a place for dualism in scientific hypothesis testing. But we pay a price when we try (consciously or unconsciously) to apply dualism in places it doesn't belong. In a radically dualistic setting science may suffer in a variety of ways when dualism is applied wholesale in places it doesn't belong to. The danger is that we will forget that dualism itself plays a role in sprouting separation, contentiousness, boundaries, bias, prejudice, and loss or lack of openness, creativity, harmony, and even ethics.

Thus, we ideally want to make use of the "power of two" without descending into two-ism.

The basic "process" of dualism is that any concept, thought, or even fact gets spun, as in a centrifuge, into its polar extreme components. The middle ground is dropped out. In radically dualistic America, for instance, we have an entrenched two-party system, with Republicans and Democrats at the poles (or liberals and conservatives). There is really no middle ground — if anyone identifies as a "moderate", we demand to know if that's a moderate republican or moderate democrat. The polar positions entrench, perceive the opposite pole as the "enemy", and build a fortress to defend — at all costs — their polar position. Without a viable middle ground — one must essentially vote or cast one's lot with one or the other extremity.

For science, this has important implications. Science in a radically dualistic environment is characterized by separations — compartmentalization of subfields of science, polarized viewpoints on every question, polarized opinions on what is considered science and what is not, and on what methods are acceptable and unacceptable for scientific study. Dualistic science assumes separation between the scientist and his object of study, which is not supported in theory (e.g., quantum theory) or in practice (e.g., Rosenthal's experimenter effects). Observation is incorrectly assumed to be objective if a few well-known sources of subjectivity are reduced (i.e., double-blind technique). Moreover, the subjectivity–objectivity dualism implies that a scientist's extra-laboratory behaviour is irrelevant to his studies, a troubling assumption at a time when ethics violations in science seem to be paralleling the rise of this in our society generally. In addition, the contentiousness between polar positions on many key scientific issues often leads to prioritizing the winning of the perceived battle against the opposition — for funding, publication, and opinion of the scientific forum. This leads to scientism, scientists grouping together to defend or uphold some particular pet theory. Conscious or unconscious data discrimination can occur due to the desire to "win" the contest, as well as due to duality's prejudicial views of the "in-group" vs. "out-group" mentality applied to different data collection methodologies.

Non-dual science offers an alternative approach which naturally addresses many of the issues with dualistic science. Non-duality is difficult to define but generally refers to a perspective or point of view in which separation does not disappear but is blended with and superseded by connectivity, relatedness, and continuity. In non-dual science, separateness is seen as an unnecessary assumption that can be dropped. Apparent separation — between people, ideas, concepts, theories, opinions — can be reframed by focusing upon similarities instead of differences, a repositioning that puts the apparent chasm of separateness into proper perspective.

For science, the implications are substantial. Non-dual science is question-focused, recreating the "middle ground" between entrenched polar positions, the "I don't know" alternative, which is the mystery from which science blooms. Non-dual science collapses objectivity and subjectivity by removing the dualistic assumption of separateness, and is therefore wholistic, incorporating both. Non-dual science respects all points of view on any issue, including dualism. Where dualism would erect boundaries and "hard" separations, non-dual science builds bridges and continua. Non-dual science is truly data-driven, and respects all forms of data for what they are, giving each form its appropriate weight in decision-making. Where dualistic science selects, segregates and excludes, non-dual science gives all forms of data a place at the table, each with its own allotment of "votes". Non-dual science is non-judgemental, relying strictly upon data for settling issues, rather than theory, presumptions or assumptions. Non-dual science focuses upon similarities rather than differences, building bridges instead of boundaries. Non-dual science is not compartmentalized, rests upon minimal assumptions, and fosters the natural development of the ethical within each scientist. Non-dual science uses all of the traditional tools of science, including dualism, but is careful to limit the use of any tool to its appropriate domain.

> Non-dual science is science, nothing more, nothing less. To return to the pure fundamental principles of scientific inquiry is to practise non-dual science, and vice versa. Moreover, non-dual science preserves and values the mystery which is the essence and *via regia* of science. Long live science, long live the scientist, and long live mystery.

Introduction

WHEN I rummage my past in search of the deeper origins of the current paper, I must go all the way back to my high school years, when I was already a scientist by natural inclination. Outside of the classroom, I eagerly read about the big bang, about new theories of the nature of light, gravitation, fluid dynamics, electromagnetism. I subscribed to *Popular Science* and *Popular Mechanics* magazines. Because I got my reading material from commercial bookstores rather than off the school library's dusty shelves, I read the "new" science of the time, the speculative stuff that did not pass peer review. At college, I majored initially in engineering physics, eventually settling instead on pure mathematics for my bachelor's degree because, frankly, it came so easy to me and was fun. I eventually discovered that it was making use of mathematics in support of the scientific enterprise which excited me most, and I found myself in graduate school specializing in statistical methods and research design.

If the story of my being born into science is rooted in high school, my awakening to the Truth about science begins in graduate school. Up until this point, I held a naïve belief in science as an open approach to investigating any question and that science was defined by the method employed to address the question, not by the content of the question itself. I was, of course, increasingly aware that there were limitations — questions which science was unable to address for methodological or ethical reasons. Moreover, at this point, the distinction between science and philosophy was not clear, the one blending seamlessly into the other where metaphysical issues were concerned.

My awakening came when, as a graduate student, I began to take an interest in a scientific field called parapsychology. I read about the studies of J.B. Rhine at the Foundation for Research on the Nature of Man in Durham, just a few miles away from UNC-Chapel Hill where I was a graduate student in behavioural statistics. I wondered whether telepathy between human beings might actually occur. It seemed to me to be a fascinating topic, one to which scientific methodology was eminently applicable, and certainly the potential payoff of such research to mankind appeared substantial. I discussed this at length with fellow graduate students and faculty and was surprised to find little awareness or interest in this area. My "awakening" came when my faculty advisor called me into his office one day and told me the faculty had unanimously agreed to request that I leave the programme. As a matriculated doctoral student for two years, they could not force me to leave but they could cut-off my funding, which they did immediately. Should I insist on staying in the programme, I was told, my funding would be restored only on condition that I denounce parapsychology, stay away from J.B. Rhine's laboratory, and withdraw from any further parapsychology activities or discussions. I would be assigned a personal "advisor" who would look over my shoulder on a month-to-month basis and sign the release for my monthly fellowship check contingent upon my steering clear of any involvement with parapsychologists or parapsychology.

At that time I didn't see the radical dualism of America as the real culprit behind this. Now I see that dualism in American society may be the distal cause of many of our sociopolitical difficulties, but that is another topic, beyond the scope of the current paper. Here, I will focus only upon the spectre of science in a radically dualistic environment.

I intend to show that while dualism may be a natural, beneficial, or useful tool in some settings, its pervasive influence in Western science is at the best a distraction and at worst a cancerous malignancy. A healthy science must be aware of these potential dangers. In this article I will explicate dualism and how I see it operating in society. Then I will present a sampling of the problems in science that may be attributable, ultimately, to dualism. There are, after all, alternatives, the best of which may be non-dualism — thus the title of the current article. I will give my description of non-dual science and outline what it can bring to the scientific enterprise.

What is Dualism?

The online *Meriam–Webster Encyclopedia* defines "dualism" as:

1: a theory that considers reality to consist of two irreducible elements or modes

2: the quality or state of being dual or of having a dual nature

3a: a doctrine that the universe is under the *dominion* of two opposing principles one of which is good and the other evil

b: a view of human beings as constituted of two irreducible elements (as matter and spirit) (http://www.merriam-webster.com/dictionary/DUALISM, accessed on 7 November 2009)

Dualism, then, is a theory, quality, state, or doctrine that chops the world into twos. Dualism is TWO-ISM. We must get beneath the skimpy and benign sounding definitional structure, though, to deepen our understanding of dualism.

Dualism, especially the *radical dualism* found in America, may be envisioned as a *process* like a centrifuge with a single bar spinning like a propeller. Every issue, idea, concept, or fact gets spun into its two extreme component polar positions.

The middle ground drops out — "middling" is not a stable position in radical dualism. Groups of devotees or adherents tend to congregate at the poles, entrench, try to recruit support, and perceive themselves in competition with the group at the opposite pole. Dualism is a very useful philosophical concept and has been put to use in a variety of ways. Dualism is the essence of the *dialectic* as described by Socrates, Hegel, and Marx, as well as in Buddhist and Talmudic sources. Dualism is a tool used in logic and is even embedded into the very heart of scientific methodology. However, like any tool, dualism has its domain of applicability, its side effects and its costs. The point being made here is that dualism can spill out of its domain of applicability into areas where it is a greater risk than benefit. We need to be aware of the problems that may be precipitated by this.

Polarization

America is *radically dualistic* because dualism is not only present but is the dominant mode of cognition affecting virtually every aspect of our lives. It is no coincidence, for instance, that America has a deeply entrenched two-party political system. Our Judeo-Christian religio-spiritual tradition is notoriously dualistic, honouring a God which is "out there" somewhere, separate from us. The human world and God's world are at polar extremes, separated from one another. America's collective mythology is dominated by the dualistic "good vs evil" theme, the endlessly repetitive struggle played over and over in our songs, stories, books, movies, and art. The combatants are polarized, the good is totally good while the bad is bad through and through. There is no middle ground, no "real" people with varying mixtures of good and bad. Many examples are easily found in the strongly polarized positions that develop on current issues such as the criminalizing of abortion, legalization of marijuana, homosexual rights, building gambling casinos, and global warming, to name just a few. People in America take sides impressively rapidly, even on issues they know little about or on which all the details are not yet known. And there is a strong distaste for the "middle ground".

No Middle Ground

Often, as when all the data is not yet available, a middle position of "I don't know", or "Let's wait and see", is the most logically appropriate. However, the polarities start forming on every issue in America long before all the details are known. The popular press and media in America constantly put forth "experts" to offer their forecasts on everything from sporting events, to Supreme Court decisions, to the president's next act and the ramifications of his last one. There is no room for a middle ground in dualism — all issues become pro vs. con, right vs. wrong, winner vs. loser. In America, if we ask people if they believe in God, for example, the domain of possible answers is simply yes or no, theist or atheist. Agnosticism, in America, is simply not a stable position. The middling response is met with, "Oh, he just hasn't made up his mind yet!" Public opinion is spun in the centrifuge of radical dualism. As has been pointed out by scholars of folklore (Claus 1998), "The American mind defines things in opposition." The benefits and advantages of a middle, moderate, or "wait-and-see" way are lost in radical dualism. Everything is black or white.

Groups Gather at the Polarities

Americans tend to define a person by his/her group memberships, including tacit "membership" in the groups that form at the poles of any issue (such as "liberal" or "conservative" political sympathies). Are you a Republican or Democrat? Having taken a stance, one becomes an ad hoc member of an advocacy group, if only mentally. This provides some social force. One then is compelled to defend their group's view (think: cognitive dissonance), and this leads to a bias or prejudice to all new information that arrives at one's doorstep. In America, we all know someone who has thrown his/her commitment to either the Democratic or Republican party and, therefore, interprets all new information according to its harmony with the party position. Anything his party does is necessarily "good" and anything the opposing party does is necessarily "bad". So polarized are we that it's difficult to find any truly independent voters any more — voters who will look cold and hard at the facts and evaluate them without pre-judgement based upon party lines.

Contentiousness, Chasms, and Hard Boundaries

An individual is either in or out of the group. From school age onwards America's radical dualism is reflected in its plethora of cliques, "in-groups", exclusive clubs and societies, membership organizations, the "A-list", and gangs that feature special clothing, styles or handshakes, and secret members — with rituals and rites to distinguish members from non-members. America is the land of the winners vs. the losers, "haves" and "have-nots", insiders vs. outsiders, activists and opposition on all issues. Members often identify themselves strongly with the group, its leadership, and ideals, sharpening the distinction between members and non-members. While the distinction between inside and outside is a natural one, in a radically dualistic society this boundary widens, hardens, and is often contentious. The boundary becomes impenetrable in the minds of the polarized participants. Here, too, there is no middle ground.

"Good" is defined idiosyncratically as membership in one's own group, with evil being the outsiders, those who look, think, speak, or behave differently than "us". If you think this sounds like racism, you're right. Racism, nationalism, ethnocentrism, and all of the other isms that separate the in-group from the out-group. It is interesting that America, which supposedly takes pride in championing individualism, has a stunning record of individuals attracted like iron fillings to the magnet of membership and conformity with the ideals of a real or tacit group at the poles of some issue.

In summary, in a radically dualistic society, every idea, concept, issue, or even fact soon becomes spun into its polarized form. People are drawn to the extremes where groups form, as in a contest. The "sides" actively recruit new members, as if outweighing the opponent and winning the seesaw is the ultimate goal. The middle ground, the range of positions between the extremes, are lost and become no-man's land. This is radical dualism at work. The goal in this report is not to debate the relative merits or alternatives to radical dualism in society, but to simply point out its most obvious effects. We'll now turn our attention to science and show how the effects noted above translate into the scientific domain, and how the scientific enterprise may be derailed by this.

Science under Dualism

Experimental science, too, is built upon a dualism — the well-worn path to experimental testing of a hypothesis. Science begins with the formulation or articulation of a hypothesis. It is actually a hypothesis pair — a dualism, or "two-ism". In a clinical trial for testing the effect of a drug, for example, we assume that the drug is either effective or not effective (with no middle ground). We envision an experiment in which some people (randomly selected!) receive the real drug and others receive something else, a control or comparison treatment. We formulate our experimental hypothesis (aka: "alternative" hypothesis) — by forecasting what should happen if the drug is effective (e.g., drug group does better than control/ comparison group). We also formulate its polar opposite, what we'd expect with an ineffective drug (e.g. drug group no better than control/comparison), and call this the "null" hypothesis.

These two statements of outcome, the pro (alternative) and the con (null) outcomes, are written in such a way as to be a complete package — either one or the other must be true, and it cannot be both. This is the essence of inductive logic that makes the system work. Thus, a hypothesis is a dualistic pair of statements specifically designed to be either/or but not both. Using this dualism, we can conduct an experimental study whose outcome will test the treatment's relative effectiveness.

But there's a sticky point, a fact that some scientists either misunderstand or simply ignore. The experimental outcome will not — cannot — reflect the treatment's effectiveness *directly* — only indirectly. There's an important logical subtlety built into this system that restricts an experiment's interpretation: one can either accept the null or reject the null, and that's all! Rejecting the null means the drug worked, the treatment group did significantly better than the comparison group. But what if your data falls into the accept the null decision box? What does that mean? You would be *wrong* if you interpret "accept the null" to mean the drug is ineffective. This only means that there's an absence of evidence to support effectiveness. As they say in the law — which is based upon the same logical dualism — "absence of evidence is not evidence of absence" (Wikipedia, http://en.wikipedia.org/wiki/Evidence_of_absence).

The benefit of this precisely prepared dualistic hypothesis statement pair, is that we can identify an effective drug with high probability, minimal false positive. The reason for this is the absence of a "middle ground" in our hypothesis statements. The "cost" of this is the uncertainty of the "accept the null" option.

As they say in the law, absence of evidence is not evidence of absence. Dualism is an excellent tool, but outside of the very restricted situation of hypothesis testing, when we move back into the world of ecologically valid interpretation and conclusion, dualism may be more of a hindrance than a help.

Boundaries and Compartmentalization of Science

It is well known that collaboration across fields is an excellent and very healthy mode of doing science. For example, in 2002, an economist (Vernon L. Smith) and psychologist (Daniel Kahneman) put their heads together to solve some knobby problems that neither could alone have done and shared a Nobel Prize for their efforts. Why isn't there more collaboration across fields? I assert that dualism is the culprit that makes this collaboration difficult. Dualism builds boundaries between the "insiders" and "outsiders". In radical dualism, those boundaries are thick, opaque, and appear impenetrable. It is natural for subfields of science to develop boundaries around themselves, so as to distinguish their work from others. As in a family group, the members have an increased commitment and trust with other members. Members of a scientific subfield are more likely to find understanding, appreciation, and sympathy for issues common to members of the subfield. However, these boundaries are temporary markers, have no solidity, and need not harden into opaqueness — it is only in a radically dualistic environment that the boundaries become impenetrable.

I can draw an example from my own work with statistical methods. Every field of science stakes a claim upon a subfield of statistical methods, such as: biostatistics, sociometrics, econometrics, psychometrics, educational statistics, engineering statistics, and so forth. Then, there is pure statistics, a subfield of mathematics. Despite the fact that statistics is statistics, and Tchebychev's Inequality is the same no matter where you apply it, the several subfields of statistics are shockingly isolated from one another. Standard textbooks preferred by these various subfields are built upon different analytic approaches (e.g., linear models vs. analysis of variance) than used in their cousin subfields, and often use different terminology for the same operations. Emphasis differs between the subfields on such issues as the appropriate place for multivariate techniques (some subfields use them widely, others virtually ban their use save in exceptional circumstances), the role of non-parametrics, and the proper use of effect size measures. Moreover, there are issues between these subfields concerning the "correct" approach to research design and data analysis, with each subfield staunchly defending and advocating its own methods as being the best, truest, and most correct. There are few advocates for the "middling" position — the view that all different methods and techniques may be simultaneously correct and valid (assuming they are based in sound mathematics) but differ in their assumptive bases and therefore present differing attractiveness for a variety of research issues to which they may be applied.

In a radically dualistic environment, therefore, science suffers from the seemingly impenetrable boundaries between subfields. In radical dualism, the tendency is to focus more upon the differences than upon the similarities, to draw distinctions rather than associations. This dualistic tendency is what ultimately keeps the biologist from consulting the social psychologist — and vice versa — on issues where the two perspectives could generate new and useful solutions for the benefit of all. Each subfield tends to develop its own variations on the proper conduct of science and may be at risk of becoming isolated from the insights and perspectives of other subfields.

"Isms" in Science

Even more disastrous is the prospect of a science that is biased and prejudiced to the point of missing out on valuable directions of research. I have discussed above how dualism splits ideas, issues, and facts into polarized pairs, making poles to attract similarly inclined individuals to form support/advocacy groups, and these groups try to recruit others to their "side" (pole), and narrow their vision on the "enemy" at the opposite pole, often going astray from their own search for truth, which is what science is supposed to be about.

In this way, dualism can be found lurking behind "scientism", the belief that science itself is the ultimate and final arbiter of all reality and knowledge and the only true "way of knowing". Huston Smith discusses scientism at length in his book, *Why Religion Matters: The Fate of the Human Spirit in an Age of Disbelief* (NY: Harper Collins, 2001). Smith expresses concern about what he sees as a growing tendency in academia towards a narrowed interpretation of truth which devalues noetic approaches to our understanding of the world we live in. Although Smith doesn't mention it specifically, there is — in my opinion — our radical dualism behind this. It is that radical dualism, instead of or in addition to the "scientism" Smith describes, that should concern us.

A classic statement of scientism, still taught in some graduate schools, is "If you can't measure it, it doesn't exist". In this statement we can see the "hard" dualistic boundary being laid.

Scepticism is another "ism" whose roots can be traced to radical dualism. Scepticism is one of the essential qualities of the great scientist. However, in recent years, in America's radical dualism, the term has been distorted and bastardized, made into something quite out of line with its original meaning. "Scepticism" has come to refer to a minor but persistent philosophical position akin to scientism. Originally, scientific scepticism meant openness to new ideas, to alternatives, to other ways of interpreting a set of findings. It meant, "let's put all the data on the table and discuss *all* of the possible explanations without pre-preferencing anyone of them". The new scepticism/scientism movement does just the opposite, reframing and using "scepticism" as a justification for squelching alternative viewpoints outside of some nebulous and ill-defined "standard" view as judged by sceptical society leadership.

This new ersatz scepticism/scientism is easily detected. The simple question to be asked in all matters of dispute in which there is less than 100 per cent agreement (such as the existence of UFO phenomena, "big foot", psychokinesis, telepathy, mediumship) is "Would you support and encourage more research on this issue?" An attitude of true scientific scepticism in its original meaning would demand a "yes" answer. The so-called "sceptic/scientism" groups reveal their true colours when they say, "No, we must stop this line of research". From their point of view, they've already got their answer. And that's my point about radical dualism — the parties at the extreme polar positions already have their answer.

The problem with this, of course, is that it only hinders the search for truth, the exact opposite of what science strives for.

Science at its best, at its most intense, is not an "ism". Science is the middle ground. The middle ground is the mystery that the great scientists have always told us was what drove them to science in the first place. Science, it can be said, is fuelled by mystery and its goal is to reduce it. Once a problem is solved, there's no more mystery and science hands it back to the engineers and marketers. Then science drops it into the "out-box" and reaches into its "in-box" for the next mystery!

Some Examples

Just such a thing is going on currently in the National Institutes of Health, National Centre for Complementary and Alternative Medicine (NIH-NCCAM). Some medical scientists want to stop funding for CAM research on the grounds that "It is inexcusable to squander scarce resources by funding unsubstantiated, non-evidence-based medical techniques that have no basis in theory or experiment" (Center for Inquiry, 2009) (http://www.centerforinquiry.net/newsroom/taxpayer_funding_for_therapeutic_touch/). This argument sounds hauntingly similar to the argument of the ersatz scepticism/scientism groups discussed above. It is basically an easy way to stop further discussion on an issue that is far from settled, except in the minds of the extreme polar groups.

Another good example of this is long-term debate during the latter part of the twentieth century concerning animal intelligence. This topic was the feature story of the October 2008 issue of *National Geographic* magazine. It is pointed out that Charles Darwin himself made observations of earthworms which so startled him that he concluded they must be "thinkers".

> Darwin went so far as to suggest that earthworms are cognitive beings because, based on his close observations, they have to make judgements about the kinds of leafy matter they use to block their tunnels. He hadn't expected to find thinking invertebrates and remarked that the hint of earthworm intelligence "has surprised me more than anything else in regard to worms".
>
> — *National Geographic*, October 2008

The article points out that the debate over animal intelligence swung in the twentieth century, first away from and later towards the positive pole. In the first part of the century the Skinnerian view of animals as no more than automatons was widespread. In support of this extreme, scientists were repeatedly cautioned concerning "anthropomorphism", and many journals scrupulously avoided terminology suggestive of it. In the latter part of the century, perspectives softened and the other extreme gained ground due to the persistence of well-known scientists such as Harvard biologist Donald Griffin.

This exemplifies the playing out of dualistic tendencies as cited in the introduction: the polarization of opinion, entrenchment, recruiting, the middle point of view dropping out, and perceiving the opposite pole as the enemy whose defeat becomes the primary goal.

Data Discrimination Due to Dualism

An important related issue is the bias and prejudice in dualistic science that unnecessarily excludes or ignores meaningful or valuable sources of data. As discussed above, radical dualism leads to in-group vs. out-group competitions and turf battles. There are many forms of data available to the scientist including quantitative and qualitative data, a variety of types of experiment, field studies, surveys, demonstrations, spontaneous and uncontrolled observations, personal reports, and anecdotes. In a radically dualistic society, the natural preference for one type of data over another type often leads to an extreme polar view of what constitutes "valid" data. For example, B.F. Skinner pioneered the *experimental analysis of behaviour* relying exclusively upon single subject designs, working intensely with a single organism (e.g. pigeon) at a time. This was innovative since multiple subject research was the mode in the human sciences at the time (save "case studies" which were common in medicine and psychology). Soon, Skinner's journal would only publish data derived from single subject design research. Thus, radical dualism led to the "hardening" of this type of data collection for Skinner's followers, no other form of data was recognized as relevant to their interests.

In the 1970s, Skinner's single subject design approach was adopted in other areas of research, such as clinical psychology (Herson & Barlowe 1976). Interestingly, in clinical studies in medicine, where single subject designs would logically be very useful, they were tacitly ignored — in favour of the multiple subject research which had become the standard in this field. This prejudicial rejection of single subject design studies in medicine is in part traceable, I believe, to dualism.

Another example of this is found in the animal intelligence debate cited above. In the 1990s several reviews of the animal cognition literature were published in journals and books. Some of them, from authors on the "con" side of the issue, concluded that there was no overall evidence for intelligence — as humans define it — in non-human species including apes and cetaceans, where numerous observations suggestive of intelligence were made. These negative reviews typically considered only controlled experimental studies and entirely excluded all field studies and anecdotal observations (through which tool use had been documented). By reviewing a small subset of the available data, only the most stringent and invasive studies, not surprisingly little evidence was found in support of animal intelligence. This strategy, conscious or unconscious on the part of the reviewers, served to defend the extreme polar view at one end of this dualism.

Objectivity vs. Subjectivity

Objectivity is one of the cornerstones of scientific observation. An objective observation is one which is independent of the bias of the observer. Without objective observation, there can be no science.

In a radically dualistic society, the concept of objective observation becomes transformed into the duality of objectivity vs. subjectivity, with no middle ground. Thus, an observation is assumed to be one or the other but not both. Under this assumption, establishing objectivity of observation is reduced to showing that commonly known subjective elements have been theoretically eliminated or reduced. For example, a scientific experiment involving human subjects is considered objective, by default, if a "double blind" is employed, presumably eliminating subjectivity. This, of course, is a gross and inaccurate misrepresentation of the complexity of subjectivity. In fact, no observation made by a human being can ever be said to be entirely free of conscious or unconscious biases. Social psychologist Robert Rosenthal and colleagues demonstrated this convincingly in a variety of experimental and educational contexts (Rosenthal 1976; Rosenthal 2002; Rosenthal & Rubin 1978). Rosenthal even showed that automating the data collection equipment does not solve the problem because ultimately it is a human being who must design what, when, how, and how often the observations will be made, and these subjective elements simply cannot be removed.

Thus, radical dualism propagates a distortion of the subjectivity/objectivity issue and is the distal cause of some of the unnecessary assumptions that scientists build into their data collection.

But scientific objectivity also has ethical implications. Objective observation, in radically dualistic settings, implies that all of the attitudes, opinions, judgements and beliefs held dear by the scientist will not influence the observation. Presumably, it does not matter whether the scientist is an environmental activist or corporate spokesperson, an advocate or critic of the research hypothesis. This implies also that a person with little or no ethical underpinning could conceivably be a productive and trustworthy scientist contributing significantly to the public knowledge base. A thief, liar, or cheater could become an expert authority on some specific scientific topic and become the go-to-guy for accurate answers on related issues.

There's evidence to show that this is fantasy. The person who comes to the laboratory to do science brings with him all of the person s/he was outside the laboratory. As an example, in 2005, the NIH identified forty-four scientists for ethics violations (Mitka, 2006). This is just the tip of the iceberg since we can be assured that there are many more ethics violators who do not get caught. Clearly, these ethics violations were behaviours that these scientists (if we can still call them that) brought with them from the outside to the inside of the laboratory — need/desire for money, career-building, egotism, etc. Certainly they have nothing to do with pure science.

In addition, Rosenthal showed that the experimental results in well-controlled formal research studies by prominent, trusted researchers, are significantly (and unintentionally) influenced by researcher variables outside of the context of the study. In short, experimental outcomes are not independent of the beliefs and expectations of the researcher(s). Thus, purely objective observation appears to be a fantasy, an ideal to be striven towards but never attainable.

In this regard, an interesting civil suit was settled in 2009 which involved this ethical issue of the separation of scientists observations from their lives outside of the laboratory (see http://www.24-7pressrelease.com/press-release/court-vindicates-doctor-who-questioned-pregnancy-miracle-report-throws-out-kwang-yul-chas-lawsuit-121781.php).

A controversial study was published a few years ago in a well-known peer-reviewed journal (Cha, Wirth, & Lobo 2001). After Bruce Flamm published articles questioning the validity of that research, first author Kwang Cha brought suit against Flamm for defamation. It was thrown out of court twice and the Court of Appeals finally affirmed that the case had no merit, upholding Flamm's contention that the science was invalid. Flamm's critique hung on the issue of the research teams' unethical behaviour outside the study. Cha had been earlier caught at plagiarism (on another publication), and second author

Daniel P. Wirth was convicted (and imprisoned) for multiple counts of fraud and identify theft in a federal court.

This interesting case suggests that — legally — the quality and purity of scientific research may be soiled if carried out by scientists whose behaviour outside the lab has been less than ethical.

To summarize, objective observation is essential to good science. However, radical dualism may distort the notion of "objective" to: (1) imply that elimination of some known sources of subjectivity (e.g., w/double blind) assures an "objective" observation (because there is no middle ground); and (2) that a scientist's general ethical outlook outside the laboratory is irrelevant to scientific pursuit. Both of these assumptions are patently false and impede the progress of science.

Summary/Conclusion of Science under Dualism

As outlined above, radical dualism spins every concept into two extreme "sides" and drops out the middle ground. Therein lays the greatest threat to science. The middle ground, the "I'm not sure" position, sprouts naturally into questions rather than contests. There's no place in science for someone who doesn't have a question that needs answering. If there's a question, there's mystery and a true search for real answers. Mystery's the key. Mystery sprouts open-mindedness, and good science blooms. Unfortunately, too much twenty-first-century science has lost its mystery, lost its openness, lost the wide-eyed curiosity that makes for great science. Too much twenty-first-century science is nothing more than marketing.

The Case for Non-dual Science

What's non-dual science? Jerry Katz has constructed a website (www.nonduality.com) on which he posts many proposed definitions of non-duality, from academic philosophers to folks who believe they are practising or experiencing non-duality.

They provide a good starting point for grasping the meaning of non-duality.

Non-duality (Advaita), literally means "not two", but Advaita does not mean "one" in the sense of eliminating all differences. The differences are present in the one in a mysterious way. They are not separated anymore, and yet they are there (Bede Griffiths).

Non-duality means nothing exists completely separate or independent of anything else . . . (even if there appears to be separation). . . .

Non-duality is an experience in which there is no separation between subject and object. . . .

In a non-dual state there is no contradiction or paradox. . . .

Non-duality is a continual correction of dualistic conceptions as they arise. It's a spontaneous process which, without judgement, playfully erases lines of division as they arise. . . .

Non-duality is total embrace, and embraces every new concept, in order to remain non-dual. . . .

Defining non-dualism is like adding legs to a snake. . . .

From these (and other) attempts at defining non-duality, we can find several threads that are repeated. First, there's an overriding sense that non-duality may be **ineffable**, impossible to define in words to everyone's satisfaction. Clearly, though, non-duality has something to do with separation, or perhaps apparent **separation which is not really separation** — a separation that results in "not two". While dualism is "two-ism", we can think of non-dualism as "not-two-ism". Non-duality can be thought of as the logical opposite of dualism. Of course, that's a paradox because it looks like a dualism again (polar opposites), but it's not because only half of the balance beam subscribes to dualism, the non-dual "pole" doesn't believe in poles! The non-dual "side" of this apparent polarity will always seek the larger perspective to see the entire centrifuge, including its middle ground, and see the entire system as the dance of life.

A third thread is that non-duality is **harmonious**, in contrast to dualism — which is seen to be strifeful. A fourth thread is that while duality deals with polarities and extremes, non-duality deals with the continuum with a broad **middle ground**. In fact, we can say metaphorically that non-duality is the middle ground, the mystery, the question and not the possible answers. Another sub-theme is that non-duality is an **attitude or perspective**, a lens through which to view the world. It is not something "out there" — in fact, non-duality doesn't award full credence to an "out there" that's different or separate from the "in here". Yet another sub-theme is that non-duality is **open, creative, unbounded, and non-judgemental**. To speak specifically about non-dual science, I have prepared the following (rough) hierarchy of concepts with commentary as a guide.

Non-dual Science is Question-focused (the Mystery)

At the heart of non-dual science is the mystery, the question, the middle ground that is dropped out of dualistic science. It is mystery that drives science but it is mystery that drives life itself. The middle ground, the "I'm not sure" position, sprouts naturally into questions rather than contests. There's no place in non-dual science for someone who doesn't have a question that needs answering. If there's a question, there's mystery and a true search for real answers. Mystery's the key. Mystery sprouts open-mindedness, and good science blooms. Unfortunately, too much twenty-first-century science has lost its mystery, lost its openness, lost the wide-eyed curiosity that makes for great science. Too much twenty-first-century science is nothing more than marketing. Non-dual science seeks to put the mystery back into science.

Non-dual Science is Wholistic (Subjective and Objective)

In non-dual science, we drop the extra assumption built into dualistic science — the assumption of distinctions, separation, and boundaries. G. Spencer Brown's remarkable attempt to construct an algebra of phenomena, *Laws of Form*, is a dualistic vision grounded in his opening assumption, "make a distinction". This distinction creates an inside and an outside, and Brown builds upon that separateness. Non-dual science drops that assumption. [**Note**: Non-dual science recognizes diversity and variation without making boundaries. In place of Brown's dualistic "make a distinction" assumption, a non-dualist version might rest upon the single assumption such as, "Choose a quality to focus upon". This leads naturally to continua.]

The consequence is that non-dual science is objective *and* subjective. Like modern quantum theory, non-dual science doesn't speak of an "inside" vs. an "outside" as Brown does. Without this assumption, the non-dual scientist brings his/her whole person to his/her science. Rather than a "distinction", the non-dual scientist recognizes the wholistic nature of himself, his life, his work, and his surroundings. He utilizes the ability to temporarily focus the mind into one or the other domain. Thus, the non-dual scientist achieves objective observation through mental focus, not through legislation. It is the same as the phenomenologists do in "shelving" their attitudes and opinions when reflecting upon an issue of choice. It is the same as the "introspection" technique used by the early psychologists (Wilhelm Wundt) to uncover "inner" data in the years before the behaviourist movement denounced inner data as invalid and focused exclusively on "outer" data (behaviour). Non-dual science, being wholistic, would attend to both inner and outer data without making "hard" distinctions.

Non-dual Science Respects all Points of View (. . . and all Possible Approaches to Solution, including Dualistic Ones, and including all Types of Data)

Non-dual science respects introspectionism and behaviourism, and the different types of data that are associated with them. It does not put (hard) boundaries between ideas, people, subfields of science, types of data or research approaches, or any of the concepts that the dualistic assumptions split into ersatz separates. In this way, non-dual science is collaborative, harmonious, open-minded, pluralistic, associationistic.

Non-dual Science is Data Driven (. . . and all data is respected)

Empirical observation is the engine of non-dual science, and all forms of observation, from the multi-subject fully randomized, controlled experimental design to the simplest of anecdotes and personal experiences are respected as "data". The diversity of data types and the diverse weightings they carry as evidence are recognized, but all data have some weight, and none are entirely excluded.

Non-dual science is a strong supporter of Occam's Razor, especially the version which is interpreted to, "recommend selection of the hypothesis that introduces the fewest assumptions and postulates the fewest entities while still sufficiently answering the question" (Wikipedia, "Occam's Razor", accessed on 25 January 2010). Moreover, non-dual science would add an important nuance to the interpretation of its meaning — the condition of harmonious agreement among ALL of the data on the issue. For example, in considering the issue of animal cognition Occam's Razor should be interpreted to demand that ALL forms of data (e.g., field observations and anecdotes) weigh in on the issue. All else being equal, the interpretation should reflect the greatest harmony among the different types of data on the issue.

Another example of this which may be even clearer is the widely publicized review of "placebo" effect by Hrobjartsson & Goetsche (2002). The review of published clinical trials that met their criteria (three-arm trials) concluded that placebo has no actual treatment effect outside of patient perceptions. However, this conclusion conflicts with decades of studies (other than three-arm) as well as clinical observation, anecdote, survey, and patient report data. Rather than discount this data entirely, the non-dual scientist would include it, albeit with appropriate weightings. Thus, by this view, Hrobartsson & Goetsche would have considered this disharmony and adjusted their conclusion accordingly.

Non-dual Science is Non-judgemental (Judgement is based on data)

In radical dualism, judgements which go beyond the available data are not only common, but they are often fiercely defended as if established truth. Non-dual scientists most certainly have opinions, pet theories, biases, and make judgements. However, the non-dual scientist tries to maintain awareness of the making of judgements and to stop or minimize this for the sake of his science. This process is reminiscent of a meditation. In meditation, the meditator understands that s/he cannot *stop* the mind from its automatic and endless production of thoughts. Instead, one raises awareness of thoughts as they arrive, and reframes, defuses or displaces them to a special area of awareness as they arrive. While the overall outcome of effects of meditation practice is relaxing and stress-reducing, meditation itself involves activity, vigilance, and alertness. So, too, the non-dual scientist does not attempt to stop the quite natural mental process of making judgements, but rather raises his awareness of when he is doing so, and seeing the judgement for what it is, holds it carefully in his consciousness in a way that it doesn't become an interference. Judgements, if allowed to bloom uninterrupted, close the door to mystery and to further discovery. Non-dual science thrives upon mystery, and so keeps judgements in check.

Non-dual Science Focuses upon Similarities (rather than differences)

The non-dual scientist does not see the boundaries and separations that dualistic science postulates (unnecessarily). Rather, s/he tends to see similarities. When faced with a critical perspective on his work, the non-dual scientist views the author of it as a person like himself with a shared interest in the mystery about which they both write. He sees the "other" as a close colleague rather than a competitor. In fact, the original meaning of "compete" reflects the non-dual view — it means to "run together", and has nothing to do with winning or losing. A colleague to run with, another opinion, a different idea, serve to provide possibilities for collaboration, for plumbing the depths of the mystery. So, too, non-dual science recognizes the similarities in the diversity of scientists, in the diversity of all human employment and activities, in the diversity of life.

Non-dual Science is Ethical (in and out of the laboratory)

The non-dual vision of relatedness and connectedness leads naturally to ethical standards — in life and laboratory. While the non-dual scientist is an individual achieving individual goals through his/her work, to the degree that he frees himself from the unnecessary assumption of separateness, he will become a part of something larger than self. His colleagues and co-workers become family of sorts, and his personal goal contributes to a larger goal. It is not important whether he thinks he can articulate what that larger goal is, or not. Being part of something larger than self, he recognizes the benefits and risks his individual presence brings to the whole. He sees the larger consequences of his activities as well.

The Non-dual Scientist Resists Compartmentalizing (being compartmentalized)

In radical dualism, there's a strong tendency to compartmentalize everything, to put it into drawers that are labelled, or files stacked in numbered file cabinets, rooms that have room numbers. In science, this extends to the compartmentalizing of subfields of study, dividing up the phenomenal universe into terse bits and assigning them to the various subfields. This is useful and beneficial. However, like the long-time counter of widgets at the widget factory who counts everything compulsively in and out of the workplace to no end whatever, simply because he "must", the scientist who sorts and catalogues everything he studies (a good and useful thing), can extend his sorting, cataloguing, and boundary-making beyond its domain of applicability. Some ideas just don't lend themselves to being divided and compartmentalized, and it may actually be counterproductive to do so with them.

Non-dual science appreciates the organization compartmentalization brings and makes full use of it when used appropriately. However, compartmentalization is focused upon differences and distinctions, and uses those distinctions to compartmentalize. Non-dual science discourages compartmentalization where it's inappropriate or inapplicable. Moreover, the non-dual scientist sees compartmentalization as a tool and like an other tool, has its range of applicability and its downsides. Non-dual science has a natural tendency in the opposite direction, to see similarities rather than differences. That is, while differences are recognized, they are not "hard" differences, leading to a sort of temporary compartmentalizing.

References

Center for Inquiry, Office of Public Policy, *A Fracture in Our Health Care:*

Paying for Non-Evidence Based Medicine. Press release. Washington, D.C.: Center for Inquiry. (Online at: http://www.center-for-inquiry.net/newsroom/taxpayer_funding_for_therapeutic_touch/.) (accessed on 28 September 2009).

Cha, K.Y., D.P. Wirth & R.A. Lobo, 2001, "Does Prayer Influence the Success of in vitro Fertilization-embryo Transfer? Report of a Masked, Randomised Trial", *J Reprod Med*. 2001 September, **46**(9): 781-87.

Claus, Peter J., 1998, "Folklore", in *India's World and U.S. Scholars, 1947-1997*, ed. Joseph W. Elder, Edward C. Dimock, Jr. and Ainslie T. Embree, New Delhi: Manohar, pp. 211-36.

Hersen, Michael and David H. Barlow, 1976, *Single-case Experimental Designs: Strategies for Studying Behavioral Change*, New York: Pergamon.

Hrobjartsson, A., Getsche P.C., 2001, "Is the Placebo Effect Powerless?: Analysis of Clinical Trials Comparing Placebo with No-treatment", *N.Engl.J.Med*, **344:** 1594–1602.

———, 2004, "Is the Placebo Effect Powerless?: Update of a Systematic Review with 52 New Randomized Trials Comparing Placebo with no Treatment", *J.Int.Med*, **256:** 91-100.

———, 2006, "Unsubstantiated Claims of Large Effects of Placebo on Pain: Serious Errors in Meta-analysis of Placebo Analgesia Mechanism Studies, *J.Clin.Epidemiol*, **59:** 336-38.

Learman, L.A., J. Avorn, D.E. Everitt & R. Rosenthal, 1990, "Pygmalion in the Nursing Home: The Effects of Caregiver Expectations on Patient Outcomes", *Journal of the American Geriatrics Society*, **38:** 797-803.

Mitka, M., 2006, "NIH ethics violations", *JAMA*, **296:** 1720.

Rosenthal, R., 2002, "Covert Communication in Classrooms, Clinics, Courtrooms, and Cubicles", *American Psychologist*, **57:** 839-49.

———, 1976, *Experimenter Effects in Behavioral Research* (Enlarged edn.). New York: Irvington.

———, 1976, "Experimenter Expectancy Effects in a Psychophysiological Experiment", *Physiological Psych.*, **4:** 137-44 (with A.M. Clarke, P.T. Michie, A.G. Andreason, & L.L. Viney).

———, 1974, "Unintended Experimenter Behaviour as Determinant of Experimental Results among Japanese Experimenters and Subjects", *Psychologia*, **17:** 140-49 (with Y. Uno, R.D. Frager, K. Takashima & A. Shibamoto).

———, 1972, "Unintended Experimenter Behavior as Evaluated by Japanese and American Observers", *J. of Soc. Psych.*, **88:** 91-106 (with Y. Uno & J. H. Koivumaki).

———, 1971, "Experimenters' Hypothesis Confirmation as Determinant of Experimental Results among Japanese Experimenters and Subjects", *Psychologia,* **14:** 24-30 (with Y. Uno, R.D. Frager & K.Takashima).

———, 1965, "Experimenter Behavior as an Unintended Determinant of Experimental Results". *J.Proj.Tech.Pers.Assess.*, **29:** 479-90 (with N. Friedman, & D. Kurland).

———, 1965, "Experimenter's Hypothesis-Confirmation and Mood as Determinants of Experimental Results", *Percept.Mot.Skills,* **20:** 1237-52 (Monogr. Suppl. 6-V20) (with P. Kohn, P. Greenfield, & N. Carota).

Rosenthal, R. & D.B. Rubin, 1978, "Interpersonal Expectancy Effects: The First 345 Studies", *The Behavioral and Brain Sciences,* **1:** 377-415.

Smith, Huston, 2001, *Why Religion Matters: The Fate of the Human Spirit in an Age of Disbelief,* NY: Harper Collins.

2

Indian Tradition, Science, and the Problem of Consciousness

Subhash Kak

Introduction

THE question of whether Indian approaches to reality can help advance the contemporary study of the phenomenon of consciousness will be taken up in this chapter.

The Vedic texts (approx. 3000–1000 BCE) claim to be *ātmavidyā* ("science of self" or "consciousness science"). The most ancient of these is the cryptic *Ṛgveda*. But prose commentaries, called the Brāhmaṇas and the Upaniṣads that appeared in the centuries following the Vedas, provide a framework to decode its narrative, establishing its central concern with consciousness.

Until recently, the question of consciousness was considered to lie outside of the scope of science and, consequently, the Indian texts on the subject were not properly examined. Scientific attitudes toward consciousness however have changed due to the recent advances in neuroscience and also because modern physics and computer science must confront the question of the observer.

In the Vedic view, reality is unitary at the deepest level since otherwise there would be chaos. This reality is called

Brahman (neuter gender). *Brahman* engenders and, paradoxically, transcends the mind/matter split. It is identical to consciousness at the cosmic scale and it informs individual minds. Turning focus to the very nature of the mind provides insight about consciousness.

Since language is linear, whereas the unfolding of the universe takes place in a multitude of dimensions, language is limited in its ability to describe reality. Because of this limitation, reality can only be experienced and never described fully. All descriptions of the universe lead to logical paradox, and *Brahman* is the category transcending all oppositions.

Vedic ritual is a symbolic retelling of this world-view. Knowledge is classified in two ways: the higher or unified and the lower or dual. The higher knowledge concerns the perceiving subject (consciousness), whereas the lower knowledge concerns objects. The higher knowledge can be arrived at only through intuition and meditation on the paradoxes of the outer world. The lower knowledge is analytical and it represents standard sciences (*śāstra*) with its many branches. In addition, *darśana* represents philosophy where the problem of self is taken together with some aspect of outer reality. There is a complementarity between the higher and the lower, each being necessary to define the other. This complementarity mirrors the one between mind and body.

According to the Vedic view, therefore, it is impossible to develop a language-based "theory of consciousness".

Quest of Consciousness in Computer Science and Physics

The quest for consciousness is indirectly the objective of achieving artificial intelligence (AI). The claim of some physicists that the universe itself is a giant machine is taken to complement the belief that true machine intelligence and self-awareness should arise after machine complexity has crossed a critical threshold. Much research in this field is speculative and it suffers from unwarranted generalizations.

Since machines only follow instructions, it is not credible that they should suddenly, on account of a greater number of connections between computing units, become endowed with self-awareness. On the other hand, if one accepts that machines will never become self-aware, one may ask why is the brain-machine conscious, whereas the silicon-computer is not? Perhaps the answer to this puzzle is that the brain is a self-organizing system which responds to the nature and quality of its interaction with the environment, whereas computers don't do that. But other ecological systems, which are biological communities that have complex interrelationship amongst its components, are self-organizing, without being self-aware. This suggests that while self-organization is a necessary pre-requisite for consciousness, it is not sufficient.

Yet another possibility is that current science, even when it considers self-organization and special structures of the brain, does not capture the essence of consciousness. The scientific framework may be incomplete in a variety of ways. We may not yet have discovered all the laws of nature, and our current theories may need major revision that have implications for our understanding of consciousness.

In truth, objective knowledge consists of many paradoxes. Accumulation of knowledge often amounts to making ad hoc choices in the underlying formal framework to conform to experience. The most fundamental, and a very ancient, antinomy is that between determinism and free will. Formal knowledge can at best be compared to a patchwork. The riddle is: How is a part of the physical world generating the individual's mental picture, in turn, creates the scientific theory that is able to describe nature so well? Why is mathematics so unreasonably effective?

Cognitive scientists and biologists have considered evolutionary aspects related to cognitive capacity, where consciousness is viewed as emerging out of language. Linguistic research on chimpanzees and bonobos has revealed that although they can be taught basic vocabulary of several hundred words, this linguistic ability does not extend to syntax. By contrast, small children acquire much larger vocabularies — and use the words far more creatively — with no overt training, suggesting that language is an innate capacity.

According to the nativist view, language ability is rooted in the biology of the brain, and our ability to use grammar and syntax is an instinct, dependent on specific modules of the brain. Therefore, we learn language as a consequence of a unique biological adaptation, and not because it is an emergent response to the problem of communication confronted by ourselves and our ancestors.

It has been suggested that human language capacities arose out of biological natural selection because they fulfil two clear criteria: an extremely complex and rich design and the absence of alternative processes capable of explaining such complexity. Other theories look at music and language arising out of sexual selection. But, howsoever imaginative and suggestive these models might be, they do not address the question of how the capacity to visualize models of world that are essential to language and consciousness first arise.

Finally, there is a philosophical critique of the search of a theory of consciousness. According to this critique, all that *normal* science can hope to achieve is a description of objects. But consciousness is a property of the subject, the experiencing "I", which, owing to its nature, for ever lies outside the pale of *normal* science. The experimenter cannot turn his gaze upon himself, and ordinary reality must have a dual aspect. This duality means that the world of objective causality is incomplete, creating a fundamental paradox: If objects are described by normal science, why is that science not rich enough to describe the body associated with the experiencing subject?

In a recent work,[1] I have considered the evidence that negates the view that the brain is an *ordinary* machine. I argue that even with self-organization and hitherto-unknown quantum characteristics one cannot explain the capacities associated with the brain. A summary of these arguments follows.

Brain and Mind

The question of consciousness is connected to the relationship between brain and mind. Reductionism takes it that they are identical, and mind is only the sum total of the activity in the brain, viewed at a suitable higher level of representation. Opposed to this is the viewpoint that although mind requires a physical structure, it ends up transcending that structure.

The mind processes signals coming into the brain to obtain its understandings in the domains of seeing, hearing, touching, and tasting using its store of memories. But a cognitive act is an active process where the selectivity of the sensors and the accompanying processing in the brain is organized based on the expectation of the cognitive task and on effort, will and intention. Intelligence is a result of the workings of numerous active cognitive agents.

The reductionist approach to artificial intelligence (AI) emerged out of an attempt to mechanize logic in the 1930s. In turn, AI and computer science influenced research in psychology and neuroscience and the view developed that a cognitive act is a logical computation. This appeared reasonable

1 S. Kak, 2004, *The Architecture of Knowledge: Quantum Mechanics, Neuroscience, Computers and Consciousness*, New Delhi: Centre for Studies in Civilizations and Motilal Banarsidass.

as long as classical computing was the only model of effective computation. But with the advent of quantum computing theory, we know that the mechanistic model of computing does not capture all the power of natural computation.

Schrödinger spoke of the *arithmetic paradox* related to the mind as being "the *many* conscious egos from whose mental experiences the *one* world is concocted". He added[2] that there are only two ways out of the number paradox:

> One way out is the multiplication of the world in Leibniz's fearful doctrine of monads: every monad to be a world by itself, no communication between them; the monad "has no windows", it is "incommunicado". That none the less they all agree with each other is called "pre-established harmony". I think there are few to whom this suggestion appeals, nay who would consider it a mitigation at all of the numerical antinomy. There is obviously only one alternative, namely the unification of minds or consciousnesses.

Languages of Description

Progress in science is reflected in a corresponding development of language. The vistas opened up by the microscope, the telescope, tomography and other sensing devices have resulted in the naming of new entities and processes.

The language for the description of the mind in scientific discourse has not kept pace with the developments in the physical sciences. The mainstream discussion has moved from the earlier dualistic models of common belief to one based on the emergence of mind from the complexity of the parallel computer-like brain processes. The two old paradigms of determinism and autonomy, expressed sometimes in terms of separation and interconnectedness, show up in various guises.

---

[2] W. Schrödinger, 1967, *What is Life? and Mind and Matter*, Cambridge: Cambridge University Press, pp. 128-29.

Which of the two of these is in favour depends on the field of research and the prevailing fashions. Although quantum theory has provided the foundation for physical sciences for seventy years, it is only recently that holistic, quantum-like operations in the brain have been considered. This fresh look has been prompted by the setbacks suffered by the various artificial intelligence projects and also by new analysis and experimental findings.

The languages used to describe the workings of the brain have been modelled after the dominant scientific paradigm of the age. The rise of mechanistic science saw the conceptualization of the mind as a machine. Although the neural network approach has had considerable success in modelling many counterintuitive illusions, there exist other processes in human and non-human cognition that appear to fall outside the scope of such models. Briefly, the classical neural network model does not provide a resolution to the question of binding of patterns: How do the neuron firings in the brain come to have specific meanings or lead to specific images?

Considering that the physical world is described at its most basic level by quantum mechanics, how can classical computational basis underlie the description of the structure (mind) that in turn is able to comprehend the universe? How can machines, based on classical logic, mimic biological computing? One may argue that ultimately the foundation on which the circuitry of classical computers is based is at its deepest level described by quantum mechanics. Nevertheless, actual computations are governed by a binary logic which is very different from the tangled computations of quantum mechanics. And since the applicability of quantum mechanics is not constrained, in principle, by size or scale, classical computers do appear to have limitations.

Why cannot a classical computer reorganize itself in

response to inputs? If it did, it will soon reach an organizational state associated with some energy minimum and will then stop responding to the environment. Once this state has been reached, the computer will now merely transform data according to its programme. In other words, a classical computer does not have the capability to be selective about its inputs. This is precisely what biological systems can do with ease.

Most proposals on considering brain function to have a quantum basis have done so by default. In short the argument is: There appears to be no resolution to the problem of the binding of patterns and there are non-local aspects to cognition; quantum behaviour has non-local characteristics; so brain behaviour might have a quantum basis.

Newer analysis has led to the understanding that one needs to consider reorganization as a primary process in the brain — this allows the brain to define the context. The signal flows now represent the processing or recognition done within the reorganized hardware. Such a change in perspective can have significant implications. Dual signalling schemes eventually need an explanation in terms of a binding field; they do not solve the basic binding problem in themselves but they do make it easier to understand the process of adaptation.

Biological Intelligence

For all computational models, the question of the emergence of intelligence is a basic one. Solving a specified problem, that often requires searching or generalization, is taken to be a sign of AI, which is assumed to have an all or none quality. But biological intelligence has gradation. Animal performance depends crucially on its normal behaviour. It may be argued that all animals are *sufficiently* intelligent because they survive in their ecological environment. Nevertheless, even in cognitive tasks of the kind normally associated with human intelligence, animals may perform well. Thus rats might find their way through a maze, or dolphins may solve logical problems or problems involving some kind of generalization. These performances could, in principle, be used to define a gradation. It has generally been assumed that the tasks that set the human apart from the machine are those that relate to abstract conceptualization best represented by language understanding. But nobody will deny that deaf-mutes, who don't have a language, do think. Language is best understood as a subset of a large repertoire of behaviour. Research has now established that animals think and are capable of learning and problem-solving.

Since non-humans do not use abstract language, their thinking is based on discrimination at a variety of levels. If such conceptualization is seen as a result of evolution, it is not necessary that this would have developed in exactly the same manner for all species. Other animals learn concepts non-verbally, so it is hard for humans, as verbal animals, to determine their concepts. It is for this reason that the pigeon has become a favourite with intelligence tests; like humans, it has a highly developed visual system, and we are therefore likely to employ similar cognitive categories. It is to be noted that pigeons and other animals are made to respond in extremely unnatural conditions in Skinner boxes of various kinds. The abilities elicited in research must be taken to be merely suggestive of the intelligence of the animals, and not the limits of it.

In a classic experiment, Herrnstein[3] presented eighty photographic slides of natural scenes to pigeons who were

---

3 R.J. Herrnstein, 1985, "Riddles of Natural Categorization", *Phil. Trans. R. Soc. Lond.*, **B308:** 29-144; R.J. Herrnstein, W. Vaughan, Jr., D.B. Mumford and S.M. Kosslyn, 1989, "Teaching Pigeons an Abstract Relational Rule: Insideness", *Perception and Psychophysics*, **46:** 56-64.

accustomed to pecking at a switch for brief access to feed. The scenes were comparable but half contained trees and the rest did not. The tree photographs had full views of single and multiple trees as well as obscure and distant views of a variety of types. The slides were shown in no particular order and the pigeons were rewarded with food if they pecked at the switch in response to a tree slide; otherwise nothing was done. Even before all the slides had been shown the pigeons were able to discriminate between the tree and the non-tree slides. To confirm that this ability, impossible for any machine to match, was not somehow learnt through the long process of evolution and hardwired into the brain of the pigeons, another experiment was designed to check the discriminating ability of pigeons with respect to fish and non-fish scenes and once again the birds had no problem doing so. Over the years it has been shown that pigeons can also distinguish: (i) oak leaves from leaves of other trees, (ii) scenes with or without bodies of water, (iii) pictures showing a particular person from others with no people or different individuals.

Other examples of animal intelligence include mynah birds who can recognize trees or people in pictures, and signal their identification by vocal utterances (words) instead of pecking at buttons, and a parrot who can answer, vocally, questions about shapes and colours of objects, even those not seen before. The intelligence of higher animals, such as apes, elephants, and dolphins is even more remarkable.

Another recent summary of this research is that of Wasserman[4]:

> [Experiments] support the conclusion that conceptualization is not unique to human beings. Neither having a human brain nor being able to use language is therefore a

---

[4] E.A. Wasserman, 1995, "The Conceptual Abilities of Pigeons", *American Scientist*, **83:** 246-55.

> precondition for cognition. . . . Complete understanding of neural activity and function must encompass the marvellous abilities of brains other than our own. If it is the business of brains to think and to learn, it should be the business of behavioural neuroscience to provide a full account of that thinking and learning in all animals — human and non-human alike.

An extremely important insight from experiments of animal intelligence is that one can attempt to define different gradations of cognitive function. It is obvious that animals are not as intelligent as humans; likewise, certain animals appear to be more intelligent than others. For example, pigeons did poorly at picking a pattern against two other identical ones, as in picking an A against two Bs. This is a very simple task for humans.

Wasserman devised an experiment to show that pigeons could be induced to amalgamate two basic categories into one broader category not defined by any obvious perceptual features. The birds were trained to sort slides into two arbitrary categories, such as category of cars and people and the category of chairs and flowers. In the second part of this experiment, the pigeons were trained to reassign one of the stimulus classes in each category to a new response key. Next, they were tested to see whether they would generalize the reassignment to the stimulus class withheld during reassignment training. It was found that the average score was 87 per cent in the case of stimuli that had been reassigned and 72 per cent in the case of stimuli that had not been reassigned. This performance, exceeding the level of chance, indicated that perceptually disparate stimuli had amalgamated into a new category. A similar experiment was performed on pre-school children. The children's score was 99 per cent for stimuli that had been reassigned and 80 per cent for stimuli that had not been reassigned. In other words, the children's performance was roughly comparable to that of pigeons. Clearly, the performance of adult humans at this task will be superior to that of children or pigeons.

Another interesting experiment related to the abstract concept of sameness. Pigeons were trained to distinguish between arrays composed of a single, repeating icon and arrays composed of sixteen different icons chosen out of a library of thirty-two icons. During training each bird encountered only sixteen of the thirty-two icons; during testing it was presented with arrays made up of the remaining sixteen icons. The average score for training stimuli was 83 per cent and the average score for testing stimuli was 71 per cent. These figures show that an abstract concept not related to the actual associations learnt during training had been internalized by the pigeon.

Animal intelligence experiments suggest that one can speak of different styles of solving AI problems. Are the cognitive capabilities of pigeons limited because their style has fundamental limitations? It is possible that the relatively low scores on the sameness test for pigeons can be explained on the basis of wide variability in performance for individual pigeons and the unnatural conditions in which the experiments are performed. But is the cognitive style of all animals similar and the difference in their cognitive capabilities arise from the differences in the sizes of their mental hardware? Since current machines do not, and cannot, use inner representations, it is right to conclude that their performance can never match that of animals. Most importantly, the generalization achieved by pigeons and other non-humans remains beyond the capability of machines.

A useful perspective on animal behaviour is its recursive nature, or part–whole hierarchy. Considering this from the bottom up, animal societies have been viewed as *superorganisms*. For example, the ants in an ant colony may be compared to cells, their castes to tissues and organs, the queen and her drones to the generative system, and the exchange of liquid food amongst the colony members to the circulation of blood and lymph. Furthermore, corresponding to morphogenesis in organisms the ant colony has sociogenesis, which consists of the processes by which the individuals undergo changes in caste and behaviour. Such recursion has been viewed all the way up to the earth itself seen as a living entity. Parenthetically, it may be asked whether the earth itself, as a living but unconscious organism, may not be viewed like the unconscious brain. Paralleling this recursion is the individual who can be viewed as a collection of several *agents* where these agents have sub-agents which are the sensory mechanisms and so on. But these agents are bound together and this binding defines consciousness.

Holistic Processing and Quantum Models

The quantum mechanical approach to the study of consciousness has an old history and the creators of quantum theory were amongst the first to suggest it. More recently, scholars have proposed specific quantum theoretic models of brain function, but there is no single model that has emerged as the favoured one at this point. In my own work I have considered the connections between quantum theory and information arguing that brain's processing is organized in a hierarchy of languages: associative at the bottom, self-organizational in the middle, and quantum at the top. Neural learning is associative and it proceeds to create necessary structures to "measure" the stimulus-space; at the higher level of multiple agents the response is by reorganizing the grosser levels of the neural structure. Each cognitive agent is an abstract quantum system. The linkages amongst the agents are regulated by an appropriate quantum field. This allows the individual at the higher levels of abstraction to initiate cognition or action, leading to active behaviour.

One striking success of the quantum models is that they provide a resolution to the determinism–free will problem. According to quantum theory, a system evolves causally until it is observed. The act of observation causes a break in the causal chain. This leads to the notion of a participatory universe.[5] Consciousness provides a break in the strict regime of causality. It would be reasonable to assume that this freedom is associated with all life. But its impact on the ongoing processes will depend on the entropy associated with the break in the causal chain.

Quantum theory defines knowledge in a relative sense. In the quantum world, it is meaningless to talk of an objective reality. Knowledge is a collection of the observations on the reductions of the wave function, brought about by measurements using different kinds of instrumentations.

The indeterminacy of quantum theory does not reside in the microworld alone. For example, Schrödinger's cat paradox transforms into microscopic uncertainty. Brain processes are not described completely by the neuron firings; one must, additionally, consider their higher order bindings, such as thoughts and abstract concepts, because they, in turn, have an influence on the neuron firings. A wavefunction describing the brain would then include variables for the higher order processes, such as abstract concepts as well. But such a definition will leave certain indeterminacy in our description.

Reorganizing Signals

Living systems are characterized by continual adaptive organization at various levels. The reorganization is a response to the complex of signal flows within the larger system. For example, the societies of ants or bees may be viewed as single superorganisms.

---

[5] J. Wheeler, 2000, *Geons, Black Holes, and Quantum Foam*, New York: W.W. Norton.

Hormones and other chemical exchanges among the members of the colony determine the ontogenies of the individuals within the colony. More pronounced than this global exchange is the activity amongst the individuals in cliques or groups.

Paralleling trophallaxis is the exchange of neurotransmitters or electrical impulses within a neural network at one level, and the integration of sensory data, language, and ideas at other levels. An illustration of this is the adaptation of somatosensory cortex to differential inputs. The cortex enlarges its representation of particular fingers when they are stimulated, and it reduces its representation when the inputs are diminished, such as by limb deafferentation.

Adaptive organization may be a general feature of neural networks and of the neocortex in particular. Biological memory and learning within the cortex may be organized adaptively. While there are many ways of achieving this, nesting among neural networks within the cortex is a key principle in self-organization and adaptation. Nested distributed networks provide a means of orchestrating bottom-up and top-down regulation of complex neural processes operating within and between many levels of structure.

There may be at least two modes of signalling that are important within a nested arrangement of distributed networks. A fast system manifests itself as spatiotemporal patterns of activation among modules of neurons. These patterns flicker and encode correlations that are the signals of the networks within the cortex. They are analogous to the hormones and chemical exchanges of the ant or bee colonies in the example mentioned earlier. In the brain, the slow mode is mediated by such processes as protein phosphorylation and synaptic plasticity. They are the counterparts of individual ontogenies in the ant or bee colonies. The slow mode is intimately linked to learning and development (i.e. ontogeny), and experience with and adaptation to the environment affect both learning and memory.

By considering the question of adaptive organization in the cortex, our approach is in accordance with the ideas of Gibson[6] who has long argued that biological processing must be seen as an active process. We make the case that nesting among cortical structures provides a framework in which active reorganization can be efficiently and easily carried out. The processes are manifest by at least two different kinds of signalling, with the consequence that the cortex is viewed as a dynamic system at many levels, including the level of brain regions. Consequently, functional anatomy, including the realization of the homunculus in the motor and sensory regions, is also dynamic. The homunculus is an evolving, and not a static representation, in this view.

It is not known how appropriate associative modules come into play in response to a stimulus. This is an important open question in neural computing. The paradigm of "active" processing in the context of memory is usually treated in one of two ways. First, the processing may be pre-set. This is generally termed "supervised learning", and it is a powerful but limited form of active processing. A second type of processing does not involve an explicit teacher, and this mechanism is termed "unsupervised learning". It is sensitive to a number of constraints, including the structure and modulation of the network under consideration.

There are different ways that biological memory may be self-organizing, and in this section, we suggest that the nesting of distributed neural networks within the neocortex is a natural candidate for encoding and transducing memory.

---

6 J.J. Gibson, 1979, *The Ecological Approach to Visual Perception*, Boston: Houghton-Mifflin.

Nesting has interesting combinatorial and computational features, and the seemingly simplistic organization of nested neural networks may have profound computational properties. However, we do not claim that nesting is the only important feature for adaptive organization in neural systems.

The neocortex is a great expanse of neural tissue that makes up the bulk of the human brain. As in all other species, the human neocortex is made up of neural building blocks. At a rudimentary level, these blocks consist of columns oriented perpendicular to the surface of the cortex. These columns may be seen as organized in the most basic form as minicolumns of about 30 μm in diameter. The minicolumns are, in turn, organized into larger columns of approximately 500–1000 μm in diameter. It has been estimated that the human neocortex contains about 600 million minicolumns and about 600,000 larger columns. Columns are defined by ontogenetic and functional criteria, and there is evidence that columns in different brain regions coalesce into functional modules. Different regions of the brain have different architectonic properties, and subtle differences in anatomy are associated with differences in function.

Beginning with cortical minicolumns, progressive levels of cortical structure consist of columns, modules, regions and systems. It is assumed that these structures evolve and adapt through the lifespan. It is also assumed that the boundaries between the clusters are plastic: they change slowly due to synaptic modifications or, more rapidly, due to synchronous activity among adjacent clusters.

Results from the study of neural circuits controlling rhythmic behaviour, such as feeding, locomotion, and respiration, show that the same network, through a process of "rewiring" can express different functional capabilities.

Superorganisms also have nested structures in terms of

individuals who interact more with certain members than others. In the case of ants, the castes provide further "modular" structure. For the case of honeybees:[7]

> [It is important to recognize] subsystems of communication, or cliques, in which the elements interact more frequently with each other than with other members of the communication system. In context, the dozen or so honeybee workers comprising the queen retinue certainly communicate more within their group (including the queen) than they do with the 100 or 200 hundred house bees receiving nectar loads from foragers returning from the field. The queen retinue forms one communication clique while the forager-receiver bees form another clique.

Another fundamental communication within the superorganism is the one that defines its constitution. This is a much slower process which can be seen, for example, when a queen ant founds her colony. The queen governs the process of caste morphogenesis.[8] Within the new colony, the queen, having just mated with her suitors and received more than 200 million sperms, shakes off her wings and digs a little nest in the ground, where she now is in a race with time to produce her worker offsprings. She raises her first brood of workers by converting her body fat and muscles into energy. She must create a perfectly balanced workforce that is the smallest possible in size, yet capable of successful foraging, so that the workers can bring food to her before she starves to death.

The queen produces the workers of the correct size for her initial survival and later, after the colony has started going, she produces a complement of workers of different sizes as

---

7 R.F.A. Moritz and E.E. Southwick, 1992, *Bees as Superorganisms*, Berlin: Springer-Verlag, p. 145.

8 M.V. Brian, 1983, *Social Insects*, London: Chapman and Hall; B. Hölldobler and E.O. Wilson, 1994, *Journey to the Ants*, Cambridge: Harvard University Press.

well as soldier ants in order to have the right organization for the survival of the colony. When researchers have removed members of a specific caste from an ongoing colony, the queen compensates for this deficit by producing more members of that caste. The communication process behind this remarkable control is not known. The communication mechanisms of the ant or the honeybee superorganisms may be supposed to have analogues in the brain.

Anomalous Abilities and Deficits Amongst Humans

That cognitive ability cannot be viewed simply as a processing of sensory information by a central intelligence extraction system, is confirmed by individuals with anomalous abilities. Idiot savants, or simply savants, who have serious developmental disability or major mental illness, perform spectacularly at certain tasks. Anomalous performance has been noted in the areas of mathematical and calendar calculations; music; art, including painting, drawing or sculpting; mechanical ability; prodigious memory (mnemonism); unusual sensory discrimination or "extrasensory" perception. The abilities of these savants and of mnemonists cannot be understood in the framework of a monolithic mind.

Oliver Sacks, in his book *The Man Who Mistook His Wife for a Hat*[9] describes two twenty-six-year-old twins, John and Michael, with IQs of 60 who are remarkable at calendrical calculations even though "they cannot do simple addition or subtraction with any accuracy, and cannot even comprehend what multiplication means". More impressive is their ability to factor numbers into primes since "primeness" is an abstract concept. Looking from an evolutionary perspective, it is hard to see that performing

[9] O. Sacks, 1985, *The Man Who Mistook His Wife for a Hat*, New York: Harper Collins.

abstract numerical calculations related to primes would provide an advantage?

The architecture of the brain provides clues to the relationships amongst the agents, and this architecture is illuminated by examining deficits in function caused by injury. One might expect aphasia to be accompanied by a general reduction in the capacity to talk, understand, read, write, as well as do mathematics and remember things. One might also suppose that the ability to read complex technical texts would be affected much more than the capacity to understand simple language and to follow commands.

In reality, the relationship between these capacities is very complex. In aphasia, many of these capacities, by themselves or in groups, can be destroyed or spared in isolation from the others. Historically, several capacities related to language have been examined.

These include fluency in conversation, repetition, comprehension of spoken language, word-finding disability, and reading disturbances.

In alexia, the subject is able to write while unable to read; in alexia combined with agraphia, the subject is unable to write or read while retaining other language faculties; in acalculia, the subject has selective difficulty in dealing with numbers.

The complex manner in which these aphasias manifest establishes that language production is a very intricate process. More specifically, it means that at least certain components of the language-functioning process operate in a yes/no fashion. These components include comprehension, production, repetition, and various abstract processes. But to view each as a separate module only tells half the story. There exist very subtle interrelationships between these capabilities which all come into operation in normal behaviour.

Blindsight and Agnosia

There are anecdotal accounts of blind people who can see some time and deaf people who can likewise hear. Some brain damaged subjects cannot consciously see an object in front of them in certain places within their field of vision, yet when asked to guess if a light had flashed in their region of blindness, the subjects guess right at a probability rate much above that of chance.

Blindsight has been explained as being a process similar to that of implicit memory or it has been proposed that consciousness is a result of a dialogue going on between different regions of the brain. When this dialogue is disrupted, even if the sensory signals do reach the brain, the person will not be aware of the stimulus.

One may consider that the injury in the brain leading to blindsight causes the vision in the stricken field to become automatic. Then through retraining it might be possible to regain the conscious experience of the images in this field. In the holistic explanation, the conscious awareness is a correlate of the activity in a complex set of regions in the brain. No region can be considered to be producing the function by itself although damage to a specific region will lead to the loss of a corresponding function.

Agnosia is a failure of recognition that is not due to impairment of the sensory input or a general intellectual impairment. A visual agnostic patient will be unable to tell what he is looking at, although it can be demonstrated that the patient can see the object. Prosopagnosic patients are neither blind nor intellectually impaired; they can interpret facial expressions and they can recognize their friends and relations by name or voice. Yet they do not recognize specific faces, not even their own in a mirror!

Prosopagnosia may be regarded as the opposite of blindsight. In blindsight there is recognition without awareness, whereas in prosopagnosia there is awareness without recognition. But there is evidence that the two syndromes have underlying similarity. Electrodermal recordings show that the prosopagnosic responds to familiar faces although without awareness of this fact. It appears, therefore, that the patient is subconsciously registering the significance of the faces. Prosopagnosia may be suppressed under conditions of associative priming. Thus if the patient is shown the picture of some other face it may trigger a recognition.

Split Brains and Unification

The two hemispheres of the brain are linked by the rich connections of the corpus callosum. The visual system is arranged so that each eye normally projects to both hemispheres. By cutting the optic-nerve crossing, the chiasm, the remaining fibres in the optic nerve transmit information to the hemisphere on the same side. Visual input to the left eye is sent only to the left hemisphere, and input to the right eye projects only to the right hemisphere. The visual areas also communicate through the corpus callosum. When these fibres are also severed, the patient is left with a split brain.

A classic experiment on cat with split brains was conducted by Ronald Myers and Roger Sperry.[10] They showed that cats with split brains did as well as normal cats when it came to learning the task of discriminating between a circle and a square in order to obtain a food reward, while wearing a patch on one eye. This showed that one half of the brain did as well at the task as both the halves in communication. When the patch was transferred to the other eye, the split-brain cats behaved different from the normal cats, indicating that their

---

10 R.E. Myers, R.W. Sperry, 1953, "Interocular Transfer of a Visual Form Discrimination Habit in Cats after Section of the Optic Chiasm and Corpus Callosum", *Anatomical Record,* **115:** 351-52.

previous learning had not been completely transferred to the other half of the brain.

Experiments on split-brain human patients raised questions related to the nature and the seat of consciousness. For example, a patient with left-hemisphere speech does not know what his right hemisphere has seen through the right eye. The information in the right brain is unavailable to the left brain and vice versa. The left brain responds to the stimulus reaching it whereas the right brain responds to its own input. Each half brain learns, remembers, and carries out planned activities. It is as if each half brain works and functions outside the conscious realm of the other. Such behaviour led Sperry to suggest that there are "two free wills in one cranial vault".

But there are other ways of looking at the situation. One may assume that the split-brain patient has lost conscious access to those cognitive functions which are regulated by the non-speech hemisphere. Or, one may say that nothing is changed as far as the awareness of the patient is considered and the cognitions of the right brain were linguistically isolated all along, even before the commissurotomy was performed. The procedure only disrupts the visual and other cognitive-processing pathways.

The patients themselves seem to support this second view. There seems to be no antagonism in the responses of the two hemispheres and the left hemisphere is able to fit the actions related to the information reaching the right hemisphere in a plausible theory. For example, consider the test where the word "pink" is flashed to the right hemisphere and the word "bottle" is flashed to the left. Several bottles of different colours and shapes are placed before the patient and he is asked to choose one. He immediately picks the pink bottle explaining that pink is a nice colour. Although the patient is not consciously aware of the right eye having seen the word "pink" he, nevertheless, "feels" that pink is the right choice for the occasion. In this sense, this behaviour is very similar to that of blindsight patients.

The brain has many modular circuits that mediate different functions. Not all of these functions are part of conscious experience. When these modules related to conscious sensations get "crosswired", this leads to synesthesia. One would expect that similar joining of other cognitions is also possible. A deliberate method of achieving such a transition from many to one is a part of some meditative traditions.

It is significant that patients with disrupted brains never claim to have anything other than a unique awareness. The reductionists opine that consciousness is nothing but the activity in the brain but this is mere semantic play which sheds no light on the problem. If shared activity was all there was to consciousness, then this would have been destroyed or multiplied by commissurotomy. Split brains should then represent two minds just as in freak births with one trunk and two heads we do have two minds.

Consciousness and Hidden Order

This chapter has considered evidence from physical and biological sciences to show how machines are deficient compared to biological systems at incorporating intelligence. To recapitulate the main points, machines fall short on two counts as compared to brains. Firstly, unlike brains, machines do not self-organize in a recursive manner. Secondly, machines are based on classical logic, whereas Nature's intelligence may depend on quantum mechanics.

The evidence from neuroscience that we reviewed showed how specific centres in the brain are dedicated to different cognitive tasks. But these centres do not merely do signal processing: each operates within the universe of its experience so that it is able to generalize individually. This generalization keeps up with new experience and is further related to other cognitive processes in the brain. It is in this manner that cognitive ability is holistic and irreducible to a mechanistic computing algorithm. Viewed differently, each agent is an apparatus that taps into the "universal field of consciousness". On the other hand, AI machines based on classical computing principles have a fixed universe of discourse so they are unable to adapt in a flexible manner to a changing universe. This is why they cannot match biological intelligence.

In my view, if machines with consciousness are created, they would be living machines, that is, variations on life forms as we know them. Since the material world is not causally closed, and consciousness influences its evolution, matter and minds complement each other. At the level of the individual, even medical science that is strongly based on the machine paradigm is acknowledging the influence of mind on body.[11]

Although we cannot construct a theory of consciousness, it should be possible to see the influence of consciousness in terms of coincidences.[12] Ancient Indian texts may not be directly useful, but the intuitions of the sages can help devise experiments to establish the reality of the consciousness principle outside the brain that is equivalent to finding a hidden order in the universe.

The idea of hidden order may be seen in the double-slit experiment in which the electrons or photons strike the screen at random locations individually but collectively they form dark and light bands. Although this is often viewed as individual particles going through both slits and interacting

---

11 R. Ader, D.L. Felten and N. Cohen (eds.), 1990, *Psychoneuro-immunology*, New York: Academic Press.

12 S. Kak, 2009, "The Universe, Quantum Physics, and Consciousness", *Journal of Cosmology* **3:** 500-10.

with themselves, an alternative explanation is in terms of hidden order in collectives.

To establish large order coincidences as hidden order, it will be essential to devise experiments in which chance is ruled out. We are speaking here of a programme somewhat different from psi, because we are not looking for consciousness as a force but rather as an ordering principle.

3

Vedānta and Neuroscience

Vinod D. Deshmukh

Abstract

Superconsciousness is a unique integrative mode of blissful conscious being (*sat-cit-ānanda*). It is serene (*praśānta*), super-attentive (*vipaśyanā*), and an undistracted peaceful presence (*prajñā*), in which an individual attains self-understanding with a clear insight and wisdom. There is a caring attention (*karuṇā*) to the present moment with freedom from unrealistic expectations, and emotional suffering. Traditionally, in Hinduism, Buddhism, and Taoism, various practices of meditation and Yoga are described for the achievement of such an extraordinary state of being. It is an inspired, creative, and altruistic way of living in the present actuality, with a continued sense of ceaseless wonderment (*āścaryam*).

Introduction

SOCIETY for Neuroscience defines the core concepts of neuroscience as follows:

1. The nervous system controls and responds to body functions and directs behaviour. The brain is the body's most complex organ,
2. Neurons communicate using both electrical and chemical signals,
3. Brain structure and functions are determined by both

genes and environment throughout life. Genetically determined circuits are the foundation of the nervous system,

4. Life experiences change the nervous sytem,
5. The brain is the foundation of mind. Intelligence arises as the brain reasons, plans, and solves problems,
6. The brain makes it possible to communicate knowledge through language,
7. The human brain endows us with a natural curiosity to understand how the world works,
8. Fundamental discoveries promote healthy living and treatment of diseases.

Vedānta is an ancient philosophy of life based on direct intuitive experiences and inspirations of exceptionally gifted and enlightened Vedic seers. These amazing insights and wisdom are recorded in the four Vedas, 108 Upaniṣads, *Brahma-Sūtra*, *Bhagavad-Gītā*, and Patañjali's *Yoga-Sūtras*. Since, Vedānta is an authentic way of life, there have been many subsequent examples of enlightened individuals, who have expressed their insights in their own words like Ādi Śaṅkarācārya's direct experience (*aparokṣānubhūti* and *ātma-bodha*), Jñāneśvara's blissful experience (*amṛtānubhava* and *jñāneśvarī*), and Rāmadāsa's *dāsa-bodha*, etc. Some of the more recent visionaries from India include Shri Ramakrishna Paramahamsa, Vivekananda, Sri Aurobindo, Raman Maharshi, Rabindranath Tagore, and J. Krishnamurti.

It is important to realize that Buddhism emerged as a reformed philosophy from the original matrix of the Vedic Hinduism. Essentially, there are more similarities than differences between the two. Neuroscience has now accepted the Buddhist way of mindfulness meditation as an effective method to achieve mental health and well-being. Significant research is being done to study the effects of meditation on various brain networks, mental functions, and human behaviour. Can we now integrate our understanding of neuroscience and Vedānta? My personal answer is "Yes".

The American Psychological Association dictionary defines the word "Superconscious" as a new term for transpersonal awareness in certain Eastern traditions with some variation in Hinduism, Buddhism, and Taoism. It is a state in which the individual attains the highest knowledge, freedom from (mental) pain (and suffering), and perfect spiritual insight, because the mind is free from (conflicting) passions and (unrealistic) desires. At their most profound level the practices of meditation and Yoga may be directed toward the ultimate achievement of such a superconscious state.

Superconsciousness is a unique integrative mode of blissful conscious being (*sat-cit-ānanda*). It is serene (*praśānta*), super-attentive (*vipaśyanā*), and an undistracted peaceful presence (*prajñā*), in which an individual attains self-understanding with a clear insight and wisdom. There is a caring attention (*karuṇā*) to the present situation with freedom from intrusive memories, unrealistic expectations, and emotional suffering. Traditionally, in Hinduism, Buddhism, and Taoism, various practices of meditation and Yoga are described for the achievement of such an extraordinary state of being and an inspired, creative, and altruistic way of living, with a continued sense of ceaseless wonderment (*āścaryam*).

Historically, superconsciousness is called (*turīya*) the fourth state of consciousness as described in the *Māṇḍūkya Upaniṣad*.[1] It is called the Fourth as it is in addition to the other three conscious states, namely deep sleep, dream sleep, and interactive wakefulness. *Turīya* or *brāhmī-sthiti* is an extraordinarily inspired state of calm, alert and intuitive

1 V.D. Deshmukh, 2004, "Turiya: The Fourth State of Consciousness and the STEP Model of Self-Consciousness, *Journal of Interdisciplinary Crossroads*, **1**(3): 551-60, December.

consciousness with completely integrated self-in-reality. It can be understood within the framework of modern neuroscience.

The author proposes that *turīya* is a unique conscious state of Non-Rapid Eye Movement Wakefulness (NREM-W) characterized by:

1. an optimal attentiveness with freedom from daydreaming and intrusive thoughts,
2. behavioural quiescence with no or minimal internal speech,
3. heightened perceptual clarity with cognitive excellence, and
4. an enhanced sense of well-being, serenity (*śānti*), compassion (*karuṇā*), and creativity (*sṛjatā*) (*fig*. 3.1).

In the current neuroscience, only three states of consciousness are recognized. The three states being (1) slow wave or non-REM sleep, (2) dream or REM sleep, and (3) daily interactive wakefulness. The fourth state of *turīya* is not

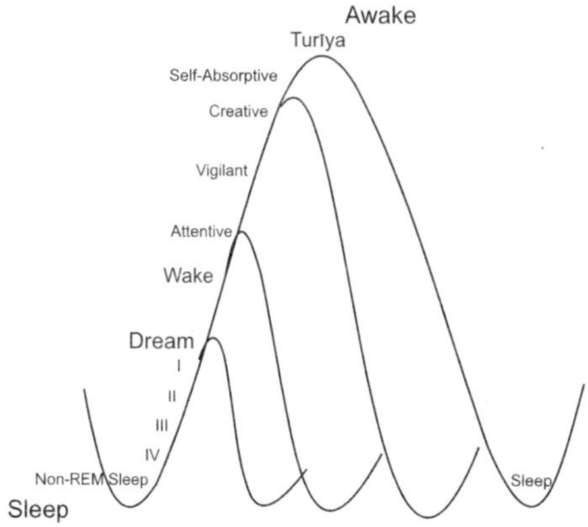

fig. 3.1: Four states of consciousness, 1. Non-REM/Deep Sleep with I-IV sub-states depending on the depth of sleep, 2. REM/dream sleep, 3. REM/wake with daydreaming and intrusive thoughts and distractions, 4. Non-REM/wake (*turīya*), with self-absorptive serenity, profound insight, and creativity.

formally recognized, but it is hinted as a meditative state with altered unitive consciousness. Some new research is being done on meditative states, especially using the functional magnetic resonance imaging (fMRI), EEG, and magnetoencephalography.[2-5]

Another important difference between the Western and Eastern perspectives regarding the conscious mind is that in the Western perspective, the mind is considered as the totality of mental and psychic activity including the psyche or the self. Whereas, in the Eastern philosophical traditions the mind has a more limited definition. In Vedānta, mind is called the inner instrument of knowledge and experience (*antaḥkaraṇa*).[6] The mind is considered to be an observable phenomenon, which becomes an "inner environment" for the essential being. During meditation, one has a feeling and conviction that this entire phenomenon is not me; this is not me, not me (*neti, neti*). I am the intrinsic and spontaneous superconsciousness.

Mind, as *antaḥkaraṇa* has four main components, namely emotional thinking (*manas*), memory (*citta*), intelligence (*buddhi*), and ego (*ahaṁkāra*). All of these components are self-observable and potentially resolvable in the ultimate enlightened matrix of *turīya*. In such an advanced state, the mental activity (*cittavr̥tti*) does not intrude automatically or habitually.

---

2 A. Lutz, J.D. Dunne, and R.J. Davidson, 2007, "Meditation and the Neuroscience of Consciousness: An Introduction", in P.D. Zelazo, M. Moscovitch and E. Thompson (eds.), *The Cambridge Handbook of Consciousness*, Cambridge: Cambridge University Press, pp. 499-551.

3 V.D. Deshmukh, 2006, "Neuroscience of Meditation", *TSW Holistic Health and Medicine*, no. 1, pp. 275-89.

4 J.H. Austin, 2009, *Selfless Insight: Zen and the Meditative Transformations of Consciousness*, Cambridge: The MIT Press.

5 J. Shear (ed.), 2006, *The Experience of Meditation: Experts Introduce the Major Traditions*, Princeton: The Infinity Foundation.

6 V.D. Deshmukh, 2009, "Introduction to Vedānta", in *Tattva* (an international online monthly magazine for Hindu Youth).

The ultimate subjectivity, self-as-subject or the essential being (*ātman*) is considered fundamental, spontaneous, intrinsic, and unobservable as a separate entity, object, process, or event. However, it is intrinsically aware of itself. It knows itself all the time. Such integrated information is the very property of a conscious living being. At its very core, there is only a sense of blissful self-existence (*svatvaṁ*) (*fig.* 3.2).

On differentiation, it manifests as the primal consciousness comprising of blissful conscious being (*sat-cit-ānanda*). On even further differentiation, the commonly recognized, three mental functions, namely conation, cognition, and affection become manifest as observable phenomena. Our sense of self is considered as an integrated, multistream reality.[7] Its four component streams, namely biophysical, mental, social and existential, integrate and function harmoniously together.

There is a considerable recent interest in the scientific study of consciousness. Several new review articles and books are

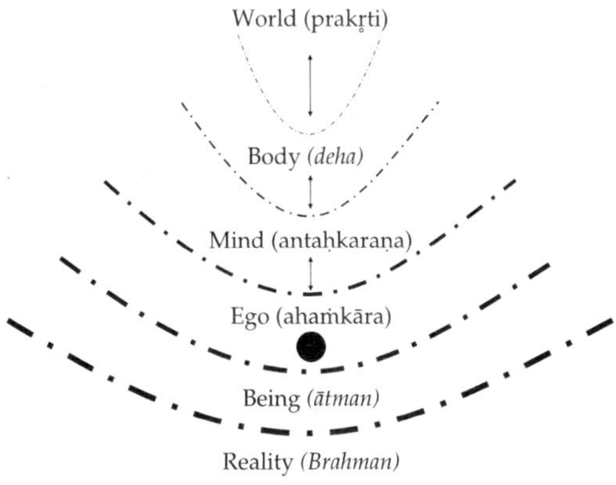

fig. 3.2: The picture of being-in-reality with nested layers of experience. The black dot in the middle is a common vantage point of ego (*jīva*). As the vantage point shifts from ego to Being/*ātman*, the mindset and self-experience changes. *Brāhmībhāva* is the realization of *Brahman* the Truth

---

[7] V.D. Deshmukh, 2008, "The Multistream Self: Biophysical, Mental, Social, and Existential", *The Scientific World Journal*, **8:** 331-41.

available to understand and address scientifically this complex topic of consciousness.[8-12] This will be discussed further below.

The Superconscious State

Historically, superconsciousness has been described as the unique fourth state of consciousness (*turīya, sthita-prajña, ātma-prasāda*, or *brāhmī-sthiti*). The four Vedas (*Ṛg, Yajur, Atharva*, and *Sāma*), which are supposed to be anonymous (*apauruṣeya*), are probably 5,000 or more years old. They have described such a state of a unique blissful conscious being.

Quotations from Vedas

> *Ātman*, the being is free of (unrealistic) desires (or fantasies) and fear of death, full of courage (calm, composure, and strength), and wisdom, self-existent, content with the essentials, lacking nothing, and the knower of itself, is healthy, cheerful, serene, and youthful.
>
> — *Atharvaveda* 10.8.44

> Realization of Being: When a person, having perceived all creatures (forms) from all directions of the world, dwells perseveringly upon the very first (primal) perception or awareness, then he/she enters into the core being by him/her. — *Yajurveda* 32.11

---

8 A. Zeman, 2002, *Consciousness: A User's Guide*, New Haven: Yale University Press.

9 B. Libet, 2004, *Mind Time: The Temporal Factor in Consciousness*, Cambridge: Harvard University Press.

10 P.D. Zelazo, M. Moscovitch and E. Thompson, 2007, *The Cambridge Handbook of Consciousness*, New York: Cambridge University Press.

11 G. Tononi and S. Laureys, 2009, "The Neurology of Consciousness: An Overview", in S. Laureys and G. Tononi (eds.), *The Neurology of Consciousness: Cognitive Neuroscience and Neuropathology*, Burlington: Elsevier, pp. 375-412.

12 C. Koch, 2009, "The Neurobiology of Consciousness", in M.S. Gazzaniga (ed.), *The Cognitive Neurosciences*, 4th edn., A Bradford Book, Cambridge: The MIT Press, pp. 1137-49.

I cannot travel to the truth of the luminous being by force (intentional effort) or duality. — *Ṛgveda* V.12.2

The Supreme Being: The enlightened person perceives within himself the Supreme Being, in whom the universe has its existence. The universe emerges to a form at the time of creation and returns to a subtle form at the time of dissolution. This Supreme Being pervades through and through all forms, like the warp and woof.

— *Yajurveda* 32.8

There is no true image of that glorious being.

— *Yajurveda* 32.3

One who knows the finite, talks excessively. One, who knows the infinite, stays quiescent. — *Yajurveda* 30.19

Quotations from Upaniṣads

The *Māṇḍūkya Upaniṣad* clearly described four states of human consciousness, namely (1) *suṣupti*, the dreamless sleep with *prajña* as the experiencing self, (2) *svapna*, the dream sleep with *taijasa* as the experiencing self, (3) *jāgrata*, the wakeful state with Vaiśvānara as the experiencing self, and (4) *turīya*, the fourth superconscious state, which is a highly awakened, self-integrated state. The primal essential being with self-awareness, in such a state, is called *ātman*, the essential self.

Māṇḍūkya Upaniṣad contains twelve original Sanskrit verses, followed by a commentary (*Kārikā*) by Gauḍapāda, who was a Sanskrit scholar and a pioneer of the non-dualistic philosophy called Advaita Vedānta. Ādi Śaṅkarācārya, who was a brilliant, eighth-century Sanskrit scholar, philosopher, and a poet of India (CE 788–820), has translated and commented on the *Māṇḍūkya Upaniṣad* as well as on Gauḍapāda's *Kārikā*. The *Māṇḍūkya Upaniṣad*, after describing the three commonly known states of human consciousness, goes on to describing *turīya* in the seventh and the twelfth verses. S. Radhakrishnan, who was a modern scholar of the Upaniṣads, translated the seventh and the twelfth verses as follows:

> (*Turīya* is) not that which cognizes the internal (objects), not that which cognizes the external (objects), not that which cognizes both of them, not a mass of cognition, not cognitive, not non-cognitive. (It is) unseen, incapable of being spoken of, ungraspable, without any distinctive marks, unthinkable, unnameable, the essence of knowledge of the oneself, that into which the world is resolved, the peaceful, the benign, the non-dual, such, they think, is the fourth quarter. He is the self; He is to be known. . . . The syllable ॐ (*auṁ*) is the very self. He who knows it thus enters the (essential) self with his (own) self.[13]

In differentiating the first state of dreamless sleep from the fully awakened fourth state of *turīya*, Gauḍapāda comments:

> in the dreamless sleep, *prajñā*, the experiencing self, is neither aware of himself nor of others. He neither knows the truth nor the untruth. He is truly not cognizant of anything. But in the state of *turīya, ātman,* the essential self is always aware of totality.

The unique conscious experience of *turīya* has been well described in *The Direct Self-Realization* (*Aparokṣānubhūti*) by Ādi Śaṅkarācārya. This experience is translated as follows:

ब्रह्मैवाहं समः शान्तः सच्चिदानन्द लक्षणः।
निर्विकारो निराकारो निरवद्यो अहम् अव्ययः।।
निरामयो निराभासो निर्विकल्पो अहम् आततः।
निर्गुणो निष्क्रियो नित्यो नित्यमुक्तो अहम् अच्युतः।।
निर्मलो निश्चलो अनंतः शुद्धो अहम् अजरो अमरः।
नाहं देहो हि असद् रूपो ज्ञानं इति उच्यते बुधैः।

brahmaivāhaṁ samaḥ śāntaḥ saccidānanda lakṣaṇaḥ ।

---

[13] S. Radhakrishnan (tr. and ed.), 1978 (1994), *The Principal Upanishads*, first published in Great Britain by George Allen & Unwin Ltd., rpt. by Indus.

nirvikāro nirākāro nirvadyo aham avyayaḥ ।।

nirāmayo nirābhāso nirvikalpo aham ātataḥ ।
nirguṇo niṣkriyo nityo nityamukto aham acyutaḥ ।।

nirmalo niścalo anantaḥ śuddho aham ajaro amaraḥ ।
nāhaṁ deho hi asad rūpo jñānaṁ iti ucyate budhaiḥ ।।

> I am truly the Absolute Being-in-Reality, which is equanimous, serene, existential, conscious, blissful, non-changing, unformed, flawless, non-dissipating, healthy, non-apparent (real), non-wavering, pervasive, attributeless, restful, enduring, steady, pure, immovable, infinite, sacred, non-ageing, imperishable. I am not merely this apparent body. This is the true Self-knowledge expressed by the wise.[14]

Patañjali's *Yoga-Sūtras* continue to be the key reference for the study of mind in Vedānta. When mind becomes calm due to practice of Yoga (योगः चित्त-वृत्ति-निरोधः *yogaḥ citta-vṛtti-nirodhaḥ*), then, the self reveals its true nature spontaneously (तदा द्रष्टुः स्वरूपे अवस्थानम् *tadā draṣṭuḥ svarūpe avasthānam*). Otherwise, the self identifies itself with the mind and its activities (वृत्ति सारूप्यं इतरत्र *vṛtti-sārūpyaṁ itaratra*).[15]

> The Self alone existed in the beginning in the shape of being, (पुरुष *puruṣa*). He, looking around, saw no one else but his self. He first uttered, "I am". Hence, he got the name of "*aham*, I". — *Bṛhadāraṇyaka Upaniṣad* I.4.1

> Let him realize that one who is omnipresent, who can't be heard, (who is) unapproachable, unthought of, uncontrolled, invisible, unknowable, and undefined listener, seer, thinker, instructor, proclaimer, knower, intelligence, and internal self (*antaryāmī*) of all created beings, is — his true self.

---

14 Swami Vimuktananda, 1966, *Aparokṣānubhūti or Self-Realization of Śaṅkarācarya*, Calcutta: Advaita Ashrama.

15 I.K. Taimani, 1961, *The Science of Yoga*, Wheaton, IL: The Theosophical Publishing House.

This *ātman* is not to be obtained by many explanations, or by intellect, or by much learning. To him, who is receptive, *ātman* reveals its true nature. — *Kaṭha Upaniṣad* I.2.23

Higher than sensorium is mind, higher than mind is intellect, higher than intellect is *jīva* (ego), higher than *jīva* is the unmanifest, higher than the unmanifest is *puruṣa* (the primal Being), who is all-pervasive, has no attributes, whom having comprehended, a man becomes liberated and fearless.

— *Kaṭha Upaniṣad* II.3.7 & 8

He can reach fast and grasp without hands or feet. He can see without eyes. He can hear without ears. He knows all that is knowable, but there is no knower of his. They speak of him as the great primal Being.

— *Śvetāśvatara Upaniṣad* III.19

There is one self-luminous being, hidden in the heart of all creatures, pervading all, the innermost self of all, the organizer of all, the witness, pure awareness without any (specifiable) qualities. — *Śvetāśvatara Upaniṣad* VI.11

He who is the eternal of all lasting, the awareness of awareness, the creator of all desires, having known that (primal) cause by self-understanding and self-discipline, one becomes free from all (psychological) limitations.

— *Śvetāśvatara Upaniṣad* VI.13

Let one know the speaker, not just the speech; the seer, not just the sights; the listener, not just the sounds; the experiencer of smells, not just the odours; the enjoyer of foods, not just tastes; the doer, not just deeds; the experiencer of pain and pleasure, not just pains and pleasures; the enjoyer of sexual union, not just sexual activity; the mover, not just movements; the thinker, not just thoughts; It is myself. It is the being. Know this.

— *Kauṣītaki Upaniṣad* III.8

Know him to be the *ātman*, ever-shining. He is free from ignorance and all bonds. He is self-existent and full of bliss. He has grown out of the idea of duality. His individual "I" has merged in the (intrinsic) consciousness itself.

The Mind in Vedānta and Neuroscience

Mind means mental activity. Mind is the product of brain activity, which depends on brain's self-organization.[16] Brain and mental activities occur concurrently like two parallel streams. Mental activity includes three classical domains of conation, cognition, and affect.

Conation means intention or volition with self-activation toward a goal. Conation is defined as the proactive, as opposed to habitual, part of motivation. Intention implies a conscious decision to perform a behaviour. Volition is the mental faculty by which an individual decides upon and commits to a particular course of action. Volition involves a sense of self and an active control over the decision, its veto power, and behaviour.

Cognition means information processing. It is the flow of information through an organism's brain and mind. Information is a reduction in uncertainty provided by a message or a stimulus. A biophysical stimulus has both energy and information. Information is knowledge about facts or ideas gained through experience and learning. Cognition includes sub-processes of perception, comprehension, remembering, reasoning, judging, imagining, and problem-solving.

The executive cognitive control is a set of processes that organize, plan, and schedule other mental operations. Cognition also means thinking or thought process, which may be verbal, visual, or tactile. Such a process is usually silent and has to be inferred from observable behaviour or verbal report. It is also symbolic in the sense that it involves experience and manipulation of recalled mental representations processed through the brain.

Affect is any experience of feeling or emotion, ranging

[16] K.M. Heilman, 2002, *Matter of Mind: A Neurologist's View of Brain-Behaviour Relationships*, New York: Oxford University Press.

from suffering to elation. Affect can be irreflexive meaning directly experienced or reflexive when the person makes the emotion an object of his or her conscious perception. Emotion is an experience with physiological and behavioural response to a personally significant event. Emotional control implies self-regulation of one's emotions and their influence on one's thoughts and behaviour. Emotional insight is an awareness of one's own and others' emotions. Emotional intelligence is the ability to process emotional information and to use it in reasoning and cognition.[17]

Feeling is a self-contained phenomenal experience. Feelings can vary along three dimensions: pleasant–unpleasant, exciting–calm, and arousal–relaxation. Approach and avoid are two fundamental emotional responses that an organism has when facing a new situation. I would add a third mode of "witnessing", namely to pause, wait, and learn about the evolving situation. Humans can experience a variety of feelings, including happiness, sadness, anger, fear, surprise, boredom, and disgust. They may also have an intuitive feeling of knowing something without any sensation or focal attention. This may happen in certain meditative states. It has been variously described as creative background, or matrix awareness.

The concept of human consciousness is under an intense investigation and debate in the neurosciences. One clear definition of consciousness is that it is a set of neuro-biological processes that allow an individual to perceive, comprehend, and act on the internal and external environments.[18]

---

[17] J. Panksepp, 1998, *Affective Neuroscience: The Foundations of Human and Animal Emotions,* New York: Oxford University Press.

[18] T.P. Bleck, 2007, "Levels of Consciousness and Attention", in C.G. Goetz (ed.), *Textbook of Clinical Neurology*, 3rd edn., Saunders-Elsevier.

Consciousness defines the present experience, and the present experience defines current consciousness. Mental activity can be conscious, preconscious, or subconscious (*fig.* 3.3).

Consciousness has three main functional components: arousal, awareness, and attention. Conscious arousal describes the degree to which an individual is able to interact with the environments. General arousal is higher in a person who is more alert to sensory stimuli, more physically active, and more reactive emotionally.[19] Conscious arousal requires the interplay between the brainstem reticular activating system and the

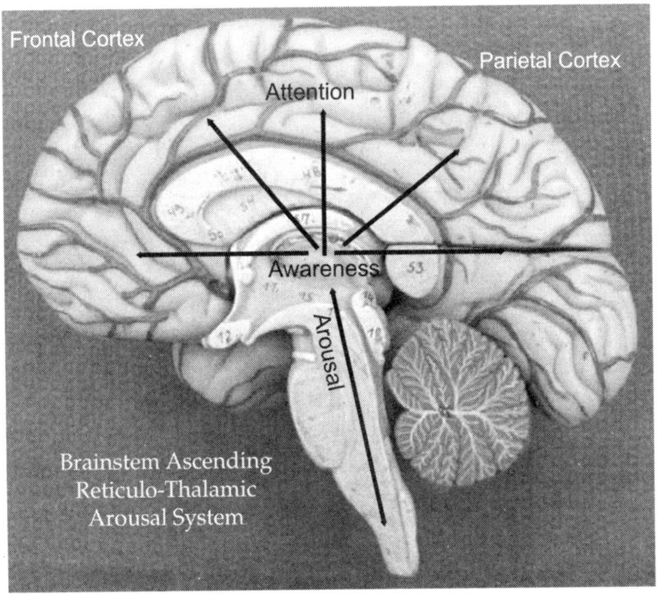

fig. 3.3: Brainstem Ascending Reticular-Thalamic Arousal system with three components of consciousness: Arousal, Awareness, and Attention. Each component has an independent, but interrelated, neural network.

---

[19] D. Pfaff, 2006, *Brain Arousal and Information Theory: Neural and Genetic Mechanisms*, Cambridge: Harvard University Press.

cerebral hemispheres. Like a pilot flame, the central brain or the centrencephalic areas are always active in all living organisms. Without it conscious life is not possible.

Conscious awareness implies that the individual is not only alert but also cognizant of self and surroundings. Awareness is global, multimodal but non-specific in terms of any particular sensory, motor, or cognitive modality. Awareness depends on arousal. The neuronal mechanism required for conscious awareness is the central thalamic nuclei with their brainstem and cortical connections.[20]

Conscious attention implies the ability to respond and process of modality-specific information at a specific time-place and person. Different sensory-motor-cognitive modalities include visual, auditory, tactile, olfactory, gustatory, locomotion, limb movements, speech, memories, thoughts, and images. Attention directed to a specific aspect of the experienced world or the self, requires both an intact general awareness and a well-functioning, modality-specific thalamo-cortical network.

The Mind in Vedānta

The Sāṁkhya system of Vedānta philosophy looks at the observable reality as an ever-changing phenomenon (*prakr̥ti*) and the ultimate, meta-conscious, transcendent being as a spontaneous and change-free person (*puruṣa*).[21] The experienced world including one's body and mind are parts of *prakr̥ti*. *Puruṣa* is the ultimate witness of the mind and

---

20 N.D. Schiff, 2009, "Recovery of Consciousness after Brain Injury: An Integrative Research Paradigm for the Cognitive Neuroscience of Consciousness", in S. Laureys and G. Tononi (eds.), *The Neurology of Consciousness: Cognitive Neuroscience and Neuropathology*, Burlington: Elsevier, pp. 1123-36.

21 Swami Prabhavananda, 1979, *The Spiritual Heritage of India*, California: Vedanta Press, Hollywood, pp. 208-25.

everything experienced (*upadraṣṭaḥ*). *Puruṣa,* the transcendent being, can never be observed as an object. *Puruṣa* is the pure intrinsic consciousness *per se* (*caitanya*). It is self-sufficient (*svārtha*). It is the living being-in-itself. There is nothing beyond it. The direct and convincing knowledge of the total reality of *puruṣa–prakṛti* is the culmination of self-knowledge (*ātmajñāna*). Then one is a light to oneself.

Vedānta considers the physical environment, the body as well as the mind as observable. Despite common belief, the mind is not our true nature. Mind is our observable, internal "environment". Mind remains active but subconscious. It is only illuminated by our spontaneous, intrinsic awareness.

The brain functions as an interface between the internal (*antara*) and the external (*bāhya*) environments. In order to survive and thrive within their natural environment, all organisms have developed effective sensory, motor, and cognitive organs for interaction. For each organism, the observed environment is continuously changing in a quantum way by the instantaneous combination of three primary forces (*guṇaḥ*), which are inertia (*tamas*), activity (*rajas*), and homeokinetic equilibrium (*sattva*). Such quantum transformation is the basis of experienced events in reality. A well-known quotation in Vedānta states that the body is an instrument for achieving life's goals (शरीरं आद्यं खलु धर्मसाधनं *śarīraṁ ādyaṁ khalu dharmasādhanaṁ*).

Vedānta names energy and information-processing organs of the body as (a) sensory-perceptual (*jñānendriya*), (b) motor-action (*karmendriya*), and (c) mental (*antarendriya*). Even our sense of self is divided into the external (*bāhyātmā*), internal (*antarātmā*) and transcendent (*paramātmā*).[22]

---

22 Swami Madhavananda, 1980, *Minor Upanishads: Angiras' Atmopanishad,* Calcutta: Advaita Ashrama, pp. 11-16.

There are five *jñānendriyas*, namely (1) eyes for visual, (2) ears for auditory, (3) nose for olfactory, (4) skin for tactile, and (5) tongue for gustatory perception and experience. The five *karmendriyas* are (1) tongue, mouth, and nose for breathing, swallowing, tasting, eating, and speech, (2) hands for skilful manual and creative work and arts, (3) feet for standing, walking, running, and playing, (4) excretory organs for elimination of bodily waste, and (5) genital organs for fulfilling sexual desires and reproductive activity. The four *antarendriyas* are: (1) *citta,* a storehouse of memory, (2) *manaḥ,* synthesis, comprehension, and analysis of perception and cognition, (3) *buddhi,* intelligent discrimination of experience and decision-making, and (4) *ahaṁkāra,* a sense of I-me-mine (*fig.* 3.4).

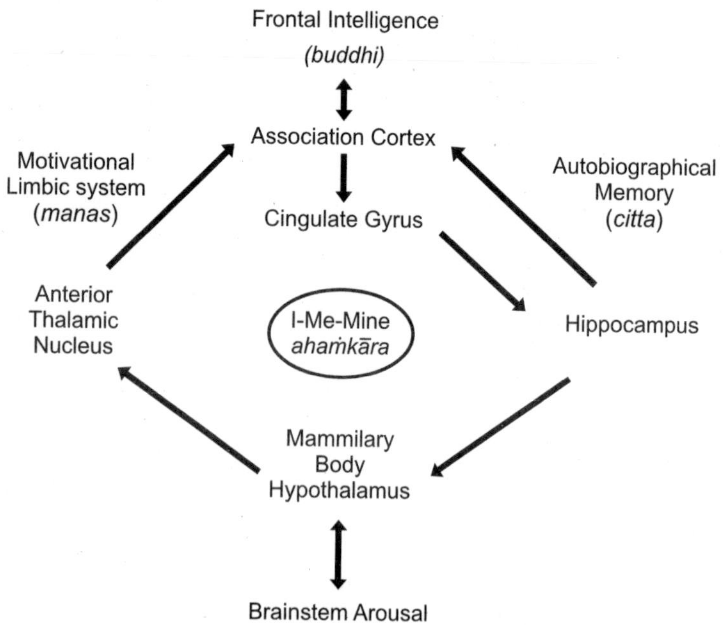

fig. 3.4: Four components of mind (*antaḥkaraṇa*): *citta, manaḥ, buddhi, ahaṁkāra*. Limbic system components include hypothalamus, anterior thalamic nucleus, orbito-frontal associative cortex, cingulate gyrus, hippocampus and amygdala

Citta is the storehouse of genetic endowment and all past memories and experiences (*saṁskāraḥ*). These *saṁskāraḥ* set the stage and context for all of our experiences and actions. Ordinarily, we are heavily conditioned and limited by these memories of the past and expectations of the future. We feel as if we are imprisoned in this self-made cage, as if we are ever committed to this life of bondage, limited understanding, and endless suffering. But in all of us there is a natural instinct for freedom. This natural desire for freedom helps us rescue ourselves from this bondage by self-awareness and learning. We can realize the final goal of complete self-freedom with an unsheakable understanding, awareness, energy, wellness, and tranquillity. That is the goal of Vedānta philosophy of life.

Manaḥ is the internal faculty of perception, comprehension, and emotional cognition. It is sometimes referred to as heart (*hṛdaya*) to include the emotional aspect of human experience and the mental eye (*manas-cakṣu*). It also includes faculties of thinking, reflection, imagination, invention, affection, desire, mood and temper. *Manas* presents the cognized information to intelligence, *buddhi*. *Buddhi* is involved in discrimination and voluntary decision-making. Therefore it is required for both knowledge and action.

Ahaṁkāra is an illusive sense of doership that arises as a result of repeated voluntary efforts, decision-making, and actions. When one realizes that most events are controlled by multiple variables including a chance factor or providence (*daivaṁ*), one's *ahaṁkāra* or sense of doership dissipates.[23]

The spontaneous meta-conscious being (*puruṣa*) is beyond these fourteen instruments (*indriyas*) of indirect (*parokṣa*) knowing. The organs of perception, cognition and action transform the non-specific awareness into a specific mental event. The conviction of the experiencer that the true being is

---

23 S. Radhakrishnan, 2004, *The Bhagavadgita*, New Delhi: Harper Collins Publishers India, 20th impression, pp. 355-56.

distinct from the experienced phenomenon is self-transcendence (*apavarga*). It is the ultimate truth within each one of us and the ultimate reality. It is the one-and-only truth, that is, *ekamevādvitīyaṁ*.

The concept of the subconscious mind (*citta*), has been elaborated in the Vedāntic literature.[24] The repository of unresolved conflicts and memories form the content of the subconscious mind. It is the arena of unfinished and repressed conflicting impulses, emotions, desires, thoughts, and ideas. There are five primary instincts: (1) self-preservation, (2) socialization, (3) sexuality and reproduction, (4) self-expression and creativity, and (5) the desire to know everything including self, others, and the world. Vedānta recommends learning and developing personal skills of self-awareness, self-analysis, and self-discipline. Such paths of growth and development of self form the basis of Yoga and meditation (*fig.* 3.5).

The subconscious or unconscious in current psychology and psychiatry consists of (1) procedural memories, like walking, swimming, riding bicycle, or tying shoe laces, (2) residual unresolved conflicts, painful, resentful, or lustful memories, and (3) contextual, recent or remote background for the present situation and self.[25]

Patañjali's *Yoga-Sūtras* continue to be the key reference for the study of mind in Vedānta. When mind becomes calm due to practice of Yoga (योग: चित्तवृत्ति निरोध: *yogaḥ citta-vṛtti-nirodhaḥ*), then, the self reveals its true nature spontaneously (तदा द्रष्टु: स्वरूपे अवस्थानम् *tadā draṣṭuḥ svarūpe avasthānam*). Otherwise, it assumes itself to be the mind with its activities (वृत्ति सारूप्यं इतरत्र *vṛtti-sārūpyaṁ itaratra*). There are five characteristic mind-

---

24 Swami Akhilanada, 1965, *Hindu Psychology: Its Meaning for the West*, London: Routledge and Kegan Paul, pp. 54-77.

25 E.R. Kandel, 2006, in *Search of Memory: The Emergence of a New Science of Mind*, NY: W.W. Norton and Company, p. 374.

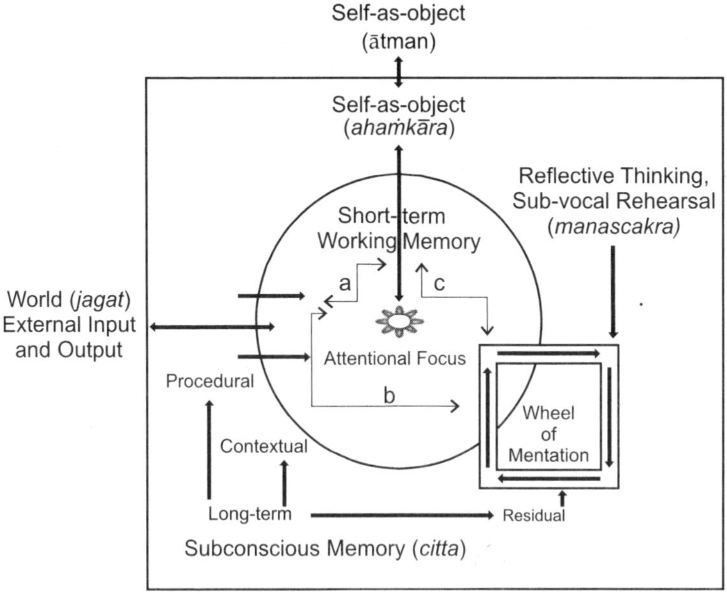

fig. 3.5: The "Black Box" of Mind: A functional model of mental activity with components of long-term memory (*citta*), activated working memory (*manas*), reflective thinking and sub-vocal rehearsal, the wheel of mentation (*manascakra*), and the self-as-object, central executive, or *ahaṁkāra* interacting with the multidimensional world (*jagat*). Three pathways of knowing are shown: (a): Direct sensory-perception, (b): Indirect, slow, reflective, memory-dependent, rumination, and (c): Spontaneous emergence of information from the subconscious to the conscious level. Total consciousness or *presence* (*ubhaya sākṣī*) includes both (a) and (c) paths simultaneously

sets: (1) Unenergetic, bored, disinterested (*mūḍha*), (2) Impulsive, unruly, chaotic (*kṣipta*), (3) Distracted (*vikṣipta*), (4) Focused, concentrated, one-pointed (*ekāgra*), and (5) Calm, composed, organized, integrated, effortless, spontaneous, self-transcendent (*niruddha*).

Patañjali's *Yoga-Sūtras* clearly define the stages of self-transformation (*samādhi* or *ātmasākṣātkāra*). *Samādhi* (enlightenment) is the actualization of one's transcendent, spontaneous being, free of all objectifiable phenomena. The successive stages of *samādhi* include freedom from four different types of contents or opacities, (1) *vitarka* is a deliberate, intentional, and self-reflected thought, (2) *vicāra* is mode of acting, proceeding, judging, thinking, and deciding, (3) *ānanda* is joy, pleasure, and happiness, (4) *asmitā* is the sense of I-me-mine.

The word *prajñā* means to know and understand clearly and fully. The word *bīja* means a seed with the potential to grow into a living being like a plant or an animal. It can also be an experiential seed of a mental state, which can grow into one's full experience. With these meanings in mind, the states of *samādhi* can be further classified into (a) with or without consciousness (*samprajñāta* or *asamprajñāta*) and (b) with or without a experiential seed (*sabīja* or *nirbīja*). The experiential seed would also indicate some content in consciousness. *Kaivalya* or *mukti* is the complete, unconditional freedom from phenomenal experience (*virām-pratyaya*) with absolute conscious clarity and an unshakeable peace and well-being, *sthita-prajña*. It is the enlightened living with a full self-realization of Truth.

The Mind in Neuroscience

Neuroscience is a scientific study of the brain and mind. A lot of progress has been made in our understanding of the brain that mediates mental processes. The brain is the organ of mental activity. It forms an interface between the internal and external environments of a living organism. In general, the internal environment is monitored and managed by the brainstem–limbic system, whereas the external environment and situations are handled by the thalamo–cortico–striate system.

The major conscious mental functions, including conation, cognition, and emotion, are experienced and responded to by the multimodal associative cortex. The back part of the associative cortex, occipito–parieto–temporal region mediates perception, cognition, attention, memory, comprehension, self-situational monitoring, and self-awareness. It has been called the semantic-conceptual field. The front part of the multimodal associative cortex cingulate–prefrontal–premotor cortex is involved in sensory–motor–cognitive and emotional integration, decision-making, working memory, intention, voluntary action, and the executive control of behaviour and mentation (*figs.* 3.6-3.7).

Most complex mental functions are mediated by cortical neural networks with multiple, reciprocal connections between different cortical areas of both cerebral hemispheres. Some areas of the brain are specialized. For instance, Wernicke's area specializes in processing language comprehension and semantic integration. Broca's area specializes in language expression including semantic, phonological, and grammatical processing. Hippocampus and the medial temporal lobe specialize in episodic memories including autobiographical memories. Right hemispheric parietal association areas specialize in attention to both right and left halves of environment and the body. Fusiform gyrus specializes in face recognition. Prefrontal cortex with its frontal intelligence is crucial for working memory, decision-making, and executive functioning, which includes initiation, continuation, pausing, and stopping of cognitive and voluntary actions[26] (*figs.* 3.8-3.9).

There are four different types of memories represented in the brain:

(1) long-term semantic memories, serving as the basis of general knowledge, are stored in the occipito-parieto-temporal association cortex,

---

[26] B.L. Miller and J.L. Cummings (eds.), 2007, *The Human Frontal Lobes: Functions and Disorders*, 2nd edn., New York: The Guildford Press.

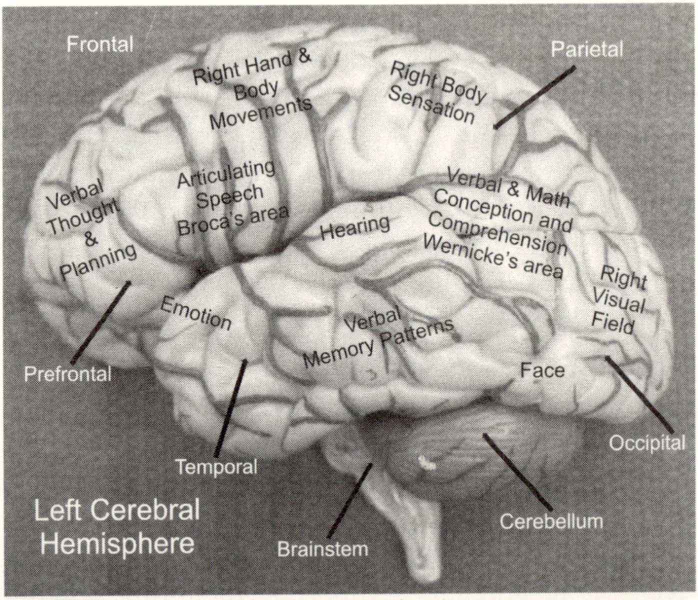

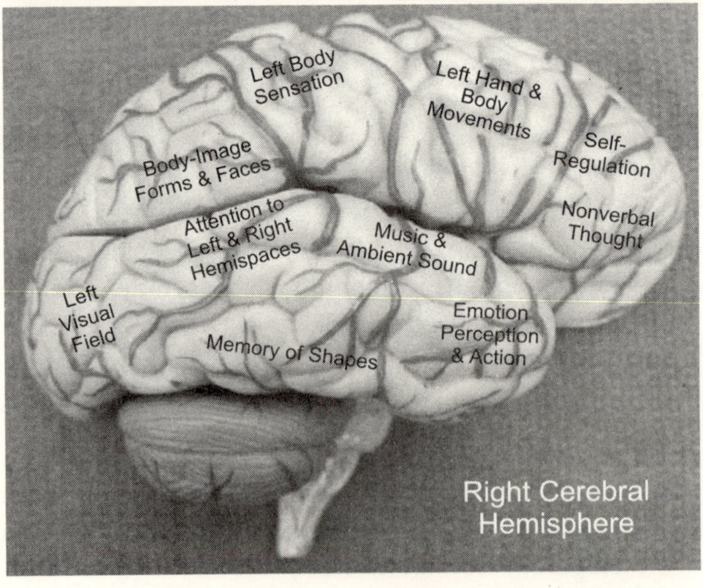

figs. 3.6-3.7: Brain model of left and right cerebral hemispheres showing various modality-specific functional areas of the cerebral cortex

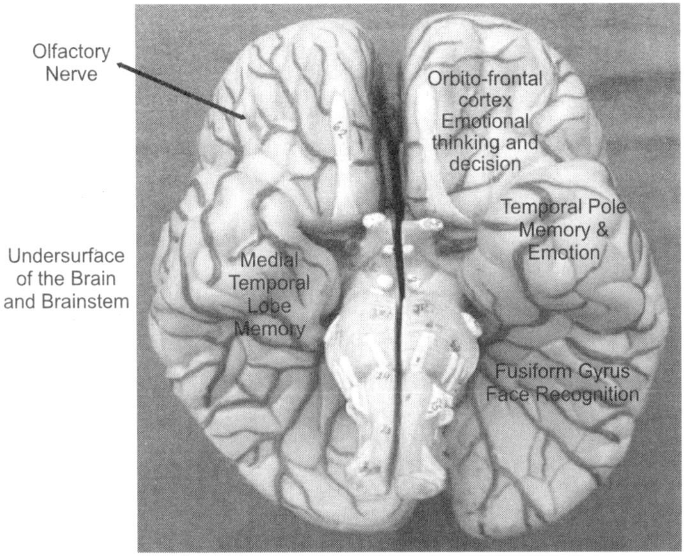

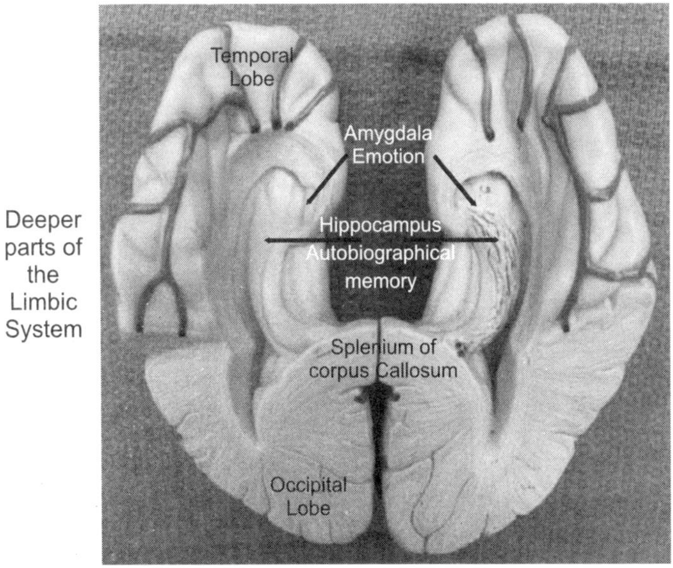

figs. 3.8-3.9: Brain model of its undersurface, and the inside structures of the temporal lobes showing hippocampus for autobiographical memory and amygdala for emotional and threat-processing

(2) episodic memories of specific events, including individual life events, are represented in the hippocampus and medial temporal lobe,

(3) procedural memory of skills and performance of learned actions is represented in frontal pre-motor cortex, basal ganglia, and cerebellum,

(4) working memory, by which one can hold and manipulate information for a brief period of time, like remembering a telephone number for 20-30 seconds, is dependent on the dorso-lateral prefrontal cortex.

The attended information in the working memory is held in active awareness by rehearsing it intentionally by internal self-talk.[27] If the information is not rehearsed, it quickly dissipates out of awareness, like the cliché, "out of sight, out of mind". This is also true of all emotions, thoughts and intentions, which dissipate if not voluntarily or habitually maintained in awareness. Like many functions of the body, the mind also follows the general principle of "dissipative Homoeostasis". So, it may be called the "homeostatic mind".

Human consciousness has three interdependent components: conscious arousal, awareness and attention.

(1) Conscious arousal depends upon a well-functioning brainstem reticular activating system. Any damage or disease in this area could result in loss of consciousness, coma, or death.

(2) Conscious awareness depends upon multimodal, non-specific thalamo-cortical system, including central thalamic nuclei and their connections. Awareness depends on optimal arousal.

---

[27] G. Ben-Yehudah and J.A. Fiez, 2007, "Development of Verbal Memory", in D. Coch, K.W. Fischer and G. Dawson (eds.) *Human Behaviour, Learning, and the Developing Brain: Typical Development*, New York: The Guildford Press, pp. 301-28.

(3) Conscious attention directs and specifies awareness and behaviour to a specific modality, time, place, and person.

Attention has two components: a global matrix function (floodlight) and a specific vector function (spotlight).[28] The global matrix function depends upon modality-non-specific thalamo-cortical circuit. The specific vector function depends upon specific thalamo–cortico–striate circuits. The attentional vector creates a specific experience with subject–object duality with a unique mental content. It can be conscious or subconscious, and voluntary or habitual.

Pre-attentive, "gist" processing can be very fast. A fearful face flashed only for 30 milliseconds (msec) can influence subsequent cognitive processing, whereas conscious attentional processing takes 300-500 msec. Pre-attentive processing is very valuable for survival in sudden threatening situations.[29]

This reminds me of Ādi Śaṅkarācārya's well-known 27th verse from *Ātma-Bodha*:

रज्जुसर्पवदात्मानम् जीवो ज्ञात्वा भयं वहेत् ।
नाहं जीवः परात्मेति ज्ञातश्चेन्निर्भयो भवेत् ।।

rajjusarpavadātmānam jīvo jñātvā bhayaṁ vahet ।
nāhaṁ jīvaḥ parātmeti jñātaścennirbhayo bhavet ।।

Having mistaken oneself as a (vulnerable mortal) being, one gets overwhelmed with fear. It is like mistaking a rope for a snake; However, when one realizes one's true infinite nature, the *ātman*, all fear goes away.

---

28 M.M. Mesulam, 1985, *Principles of Behavioural Neurology*, Philadelphia: F.A. Davis Company, pp. 125-68.

29 J.E. LeDoux, D. Schiller and C. Cain, 2009, "Emotional Reaction and Action: From Threat Processing to Goal-Directed Behaviour", in S. Laureys and G. Tononi (eds.) *The Neurology of Consciousness: Cognitive Neuroscience and Neuropathology*, Burlington: Elsevier, pp. 905-24.

In the specific cortical areas, aspects of an experienced object and the organism's ongoing response are integrated. For instance, the ventral occipito–temporal pathway, the ventral attentional system, mediates what the object is; the dorsal occipito–parietal pathway, the dorsal attentional system, is involved in processing where the object is; the prefrontal–cingulate–parietal cortex processes when to respond, or not to respond.

From moment to moment, each organism has to make a decision either to approach or to avoid the new object or situation. Frontal lobes process the avoidance response, whereas parietal lobes are involved in the approach response. Ideally, every organism has to be ever-prepared or ready, ever mentally clear and composed, and in full self-control with the ability to allow, pause, or stop one's action, if necessary. Conscious preparedness depends on emotional arousal; conscious clarity depends on spacious working memory; and full cognitive-control depends on intelligent activational and inhibitory capacities. These are the skills that we have to learn to live a good life. Yoga is skilful living (योगः कर्मसु कौशलं। *yogaḥ karmasu kauśalaṁ*).

Ipseity, Ātman, Consciousness and Presence

In a long tradition, Vedic scholars contemplated on the self (*ātman*), understood its reality and expressed it in an amazingly precise way. The Vedic literature is ripe with great insights into our true nature, the *ātman* and the nature of the ultimate universal reality (*Brahman*). These visionaries concluded that *ātman* and *Brahman* are essentially the same. This great Vedāntic philosophy of life has been a constant source of inspiration, understanding, joy, courage, and altruism for many generations in India and the world. Hopefully, it will continue to be so for many future generations of humanity.

In modern neuroscience and psychiatry, a new model of

consciousness and self-consciousness has been proposed.[30] In this model, consciousness is divided into phenomenal and introspective consciousness. The phenomenal consciousness is a first-person-perspective, whereas introspective, reflective consciousness is a second-person-perspective. The contents of phenomenal consciousness are pre-reflexive, raw feelings or qualia. They may be emotional, sensory, or memory-related. These experiences may be conscious, pre-conscious, or subconscious, depending upon the level of awareness directed to them.

The sense of self-consciousness is divided into two components, ipseity and self-qualia. Ipseity (Ipse: I or self) is the unifying basic tone of the first-person's experience. It is the centre of all conscious experience and is dependent upon a continuous, internally generated, interoceptive brain input. It gives us a sense of a single unitive being. Ipseity or I, in every experience is implicitly and pre-reflexively present in the unified field of awareness, through which all conscious experiences pass. It is an intrinsic property of the essential awareness (*fig.* 3.10).

Ipseity is a spontaneous neurobiological phenomenon, which depends upon brain arousal and awareness. Ipseity itself is non-interactive and not a pole until one starts interacting with the "other" by reflective consciousness. Self-qualia are responsible for the feelings of self-agency, self-unity, self-affectability, and implicit self-history. Ipseity is an unreflected presence with self-luminosity (*svayam jyotiḥ*).

From our common experience, we know that (i) self is distinct from situation and the environment, (ii) it is the source or author of our actions and thoughts, and (iii) it is the same person across our lifetime. The contents of self-qualia, when

---

30 T. Kircher and A. David, 2003, *The Self in Neuroscience and Psychiatry*, Cambridge: Cambridge University Press, pp. 445-75.

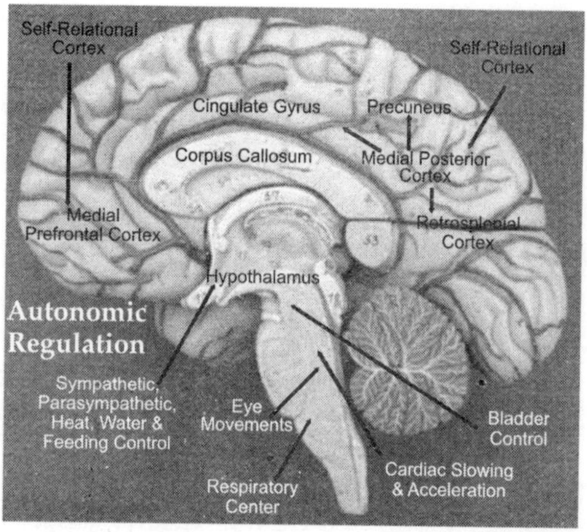

fig. 3.10: Autonomic and self-related regulatory activities. The self-relational cortex, also called the default mode network, seems to be always active metabolically, even during resting states, and its activity decreases during goal-directed task performance. See Raichle and Snyder[31]

reflected upon, become the contents of reflective consciousness. With reflective consciousness, we can communicate verbally and non-verbally with others and ourselves.

The concept of presence refers to a pre-reflexive first-person experience, i.e. ipseity and a sense of immersion (absorption) in the surrounding world.[32] Subject and object are two abstract poles of a unique structure, which is presence.

The concept of presence was discussed in greater detail by the author.[33] Presence is the fact or state of being present.

31 M.E. Raichle and A.Z. Snyder, 2009, "Intrinsic Brain Activity and Consciousness", in S. Laureys and G. Tononi (eds.) *The Neurology of Consciousness: Cognitive Neuroscience and Neuropathology*, Burlington: Elsevier, pp. 81-88.

32 J. Parnas, 2003, "Self and Schizophrenia: A Phenomenological Perspective", in *Essential Sources in the Scientific Study of Consciousness*, Cambridge: The MIT Press, pp. 217-41.

33 V.D. Deshmukh, 1990, *Presence: The Key to Mental Excellence,* pp. 119-21.

It means being present-minded as opposed to being absent-minded, which is "absence". Presence is being attentive to what is happening now and here. Presence is being in existence now. Presence is functioning fully or optimally at the present time. It implies freedom from the burden of the past and imaginations of the future. It implies renewing oneself from moment to moment. Presence is being well adapted to one's situation and self. Presence is the process of reducing uncertainty, a process of knowing and doing. Presence heightens awareness and may lead to intuitive consciousness generating new ideas, understanding, and solutions. Presence is a way to achieve mental clarity and peace. It is the very core or essence of our being.

Brain and Consciousness

The term consciousness is difficult to define, although all of us understand what it is from our first-person experience. There are many definitions and several reviews on the subject of consciousness. In clinical neurology, consciousness is defined as a set of neural processes that allow an individual to perceive, comprehend, and act on the internal and external environments.[34] It has three-functional aspects, arousal, awareness and attention. For conscious awareness, proper functioning of the cerebral cortex is essential, but it may not be directed and modality-specific. Attention depends both on awareness and arousal, and implies the ability to direct and to respond to modality-specific stimuli. It implies directed awareness and response. It is modality-specific, e.g. visual or auditory, and a precise left or right-directed action or verbal response. The right cerebral hemisphere is dominant for attentive behaviour. It is also dominant for visio-spatial and

---

[34] T. Bleck, 2007, "Levels of Consciousness and Attention", in C.G. Goetz (ed.) *Textbook of Clinical Neurology*, 3rd edn., Saunders-Elsevier, pp. 3-19.

musical abilities. The left cerebral hemisphere is usually dominant for language and symbolic processing such as skills in mathematics.[35]

Consciousness is viewed as a self-generating, self-organizing process from an ecological perspective.[36] The evolution of consciousness is seen in the context of energy-driven evolution in general, where energy and information are understood as two sides of the same coin. From this perspective, consciousness is viewed as an ecological system in which streams of perceptual, cognitive, and emotional information form a rich matrix of interactions, analogous to the interactive metabolism of a living cell. The result is an organic, self-generating or "autopoietic" system continuously in the act of creating itself. Evidence suggests that this process is non-linear, chaotic, or at least chaotic-like and capable of assuming a number of distinct states best understood as chaotic attractors.

There are two properties of consciousness, integration and differentiation. Conscious experience is integrated; each conscious scene is unified, and at the same time, it can be highly differentiated; a huge number of different conscious states are possible at each conscious moment. Each conscious state comprises of a single scene that is private and cannot be decomposed into independent components. We cannot be aware of two incongruent scenes simultaneously. So also, we cannot make more than a single conscious decision within an interval of a few hundred milliseconds (300-500 msec), the so-called psychological refractory period. Conscious experience depends upon activation or deactivation of specific neocortical

---

35 B.J. Baars, W.P. Banks and J.B. Newman (eds.), 2003, *Essential Sources in the Scientific Study of Consciousness*, Cambridge: The MIT Press.

36 S. Goerner and A. Combs, 1998, "Consciousness as a Self-Organizing Process: An Ecological Perspective", *Bio-Systems*, **46:** 123-27.

areas, in response to external or internal stimulation and activation by memory, intention, imagery or dreams (*figs.* 3.11 and 3.12).

The process of human waking from deep sleep has been studied by using PET scanning.[37] After waking from stage II of deep sleep, brain blood flow is rapidly re-established in the centrencephalic regions (upper brainstem, central thalamus, and related cortex), suggesting that the reactivation of these regions is essential for the re-establishment of basic wakeful awareness. In the next fifteen minutes, there is a further increase in the brain blood flow in the prefrontal and anterior cingulate cortex, coinciding with the re-establishment of functional alertness and the dissipation of sleep inertia.

Brain and the Sense of Self

The word "self" has multiple meanings. The dictionary defines self as (i) the total, essential or particular being of one person, the individual, (ii) personality, character, individuality, and (iii) individual's consciousness of his own being, identity, subjectivity, ego. Self-consciousness implies conscious awareness of oneself, one's own being, one's actions, thoughts and emotions.

In psychotherapy, it is acknowledged that self is a very personal matter and can never be known in its entirety. It is also difficult to understand and to define.[38] However, self is defined tentatively as the totality of a complex, dynamic and organized system of learned beliefs that an individual holds to be true about his or her personal existence. It provides a

---

[37] T.J. Balkin et al., 2002, "The Process of Awakening: A PET Study of Regional Brain Activity Patterns Mediating the Re-establishment of Alertness and Consciousness", *Brain*, **125:** 2308-19.

[38] W.W. Purkey and P.H. Stanley, 2002, "The Self in Psychotherapy", in D.J. Cain and J. Seeman (eds.) *Humanistic Psychotherapies*, Washington, D.C.: American Psychological Association.

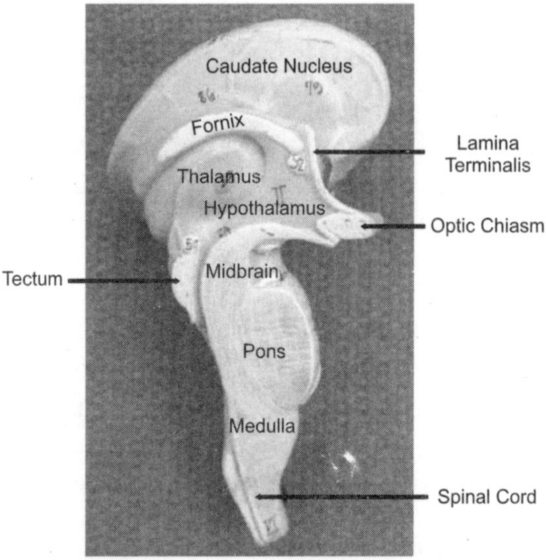

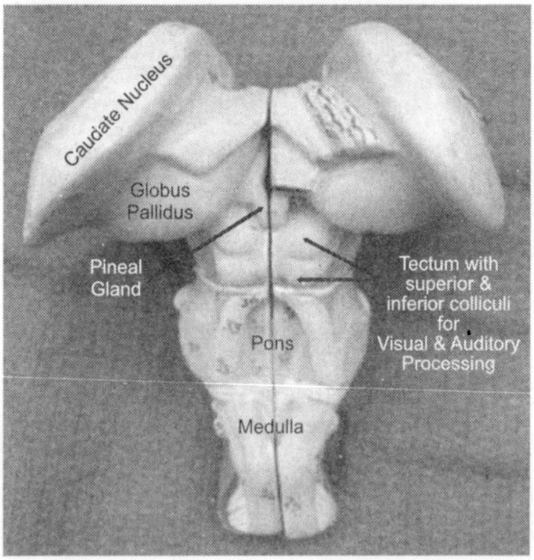

figs. 3.11 and 3.12: Brainstem and Basal Ganglia, inside and back views showing the key functional structures.

consistency to the human personality. The global self is orchestrated and balanced like a spiral, and it is centred on the "I", the self-as-subject. This "I" is the living connection with human experience. In addition to the "I" as the subject, the self also contains many self-images based on self-perception, as the self-as-object, the so-called "me".

The self is modifiable in the sense that new perceptions are integrated into the self-image throughout life, while old perceptions fade away. This continuous perceptual–cognitive flow and assimilation of experiences create flexibility and modifiability of the self. A healthy self is highly flexible, well adjusted, creative and adaptive. It is continuously involved in creating new visions of personal growth and enhancement (*fig.* 3.13).

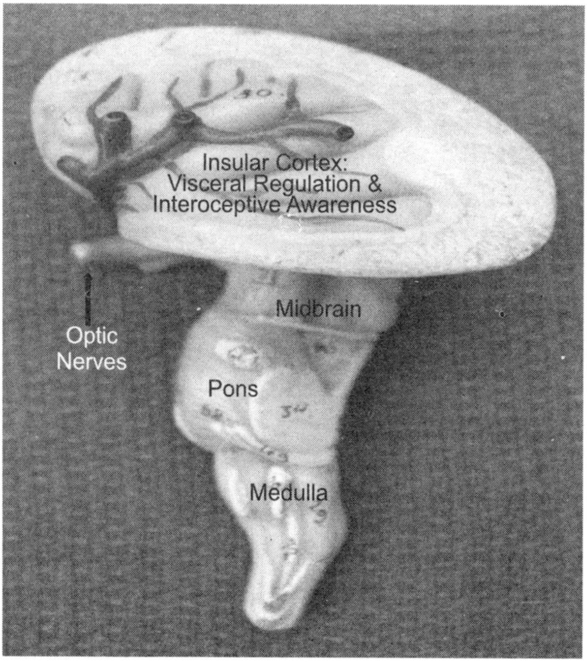

fig. 3.13: Side view of the left insular cortex and brainstem. Insular cortex processes interoceptive information and sub-serves biophysical self-awareness and regulation.

Damasio divided the sense of self into (1) proto-self, (2) core self, and (3) autobiographical self.[39]

1. Proto-self is defined as a coherent first-order collection of neural patterns, which map, moment by moment, the state of the physical structure of the organism in its many dimensions, at multiple neural levels, from brainstem to cerebral cortex. Proto-self does not perceive or hold knowledge. It has no language and we are not conscious of proto-self. It is a reference point in consciousness but it is not an interpreter of anything.
2. The core self is a transient, second-order, non-verbal account that occurs whenever an object, a sensory, emotional, or memory input, modifies the proto-self. Core consciousness is created in pulses, each pulse triggered by each objective input that we interact with or recall. We are conscious of the core self. The mechanism that produces the core self, undergoes minimal change across a lifetime. The core self requires the presence of the proto-self.
3. The autobiographical self requires the presence of the core self. It is based on the memory of multiple individual experiences of the past and the anticipated future. The autobiographical memory grows with life experiences and gets modified continuously by our new experiences. We are conscious of the autobiographical self.

Self-awareness is our capacity to sense our being. Such capacity depends on the proper functioning of medial prefrontal cortex as well as posterior cingulate, precuneus, and retro-splenial cortices. The frontal lobes, especially the

---

39 A.R. Damasio, 1999, *The Feelings of What Happens*, NY: Hartcourt Brace and Company.

prefrontal areas, specialize in maintaining working-memory representations that integrate verbal and spatial information.[40] The non-dominant frontal lobe seems to be particularly important for accessing, integrating, and maintaining self-knowledge as seen in patients with fronto-temporal dementia.[41]

The prefrontal cortex is devoted to the organization of action in all neurobiological and cognitive domains: skeletal movement, eye movement, speech, and logical reasoning.[42] Prefrontal cortex is in-charge of the temporal organization of behaviour, speech, and thinking. Three cognitive operations essential to serve these functions are: (1) preparatory set, (2) working memory, and (3) inhibitory control of interference. Prefrontal cortex is able to sculpt behaviour and perform diverse tasks through parallel excitatory and inhibitory regulation of neural activity in distributed neural networks.[43]

Llinas considers "prediction" as the ultimate function of the brain.

> Self is the centralization of prediction. The self is not born out of consciousness, only the noticing of it (i.e. self-awareness). According to this view, the self can exist without awareness of its own existence. Even for us as self-aware individuals, self-awareness is not continuously present. . . .
>
> "I" has always been a magnificent mystery; I believe, I say, I whatever. But one must understand that there is no such tangible thing. It is just a particular mental state, a generated

---

40 V. Prabhakaran, K. Narayanan, Z. Zhao and J.D.E. Gabriel, 2000, "Integration of Diverse Information in Working Memory within the Frontal Lobe", *Nature Neuroscience*, **3**(1): 85-90, January.

41 E.K. Miller and J.D. Cohen, 2001, "An Integrative Theory of Prefrontal Cortex Function", *Annu. Rev. Neurosci*, **24**: 167-202.

42 J.M. Fuster, 1999, "Synopsis of Function and Dysfunction of Frontal Lobe", *Acta Psychiatr Scand* **99** (suppl. 395): 51-57.

43 D.T. Stuss and R.T. Knight (eds.), 2002, *Principles of Frontal Lobe Function*, Oxford University Press.

> abstract entity we refer to as "I" or "self". . . . The "I" of the vortex, that which we work for and suffer for, is just a convenient word that stands for as global an event as does the concept of Uncle Sam *vis-à-vis* the reality of a complex, heterogeneous United States.[44]

Creative innovation has been recently defined as the ability to understand and express novel orderly relationships. It can help us to help each other and our humanity to flourish in harmony with Nature. Nature is the unfathomable wonder, where we find ourselves in. There is always more to learn and discover about ourselves and nature, throughout our precious, but limited conscious life.

The Multistream Nature of Self

I am proposing to extend the standard bio–psycho–social model of mind and self to a more comprehensive concept of bio–psycho–social–existential self and mind. By adding the existential component to the self-concept and mind, the total human experience is better understood and explained. The word "existential" refers to the concrete lived experience of human beings, firmly rooted in the real world. Existential living is the capacity to live fully in the present, and respond freely and flexibly to new experience without fear. Such integrative living would create an experience of presence with an awareness and understanding of one's current existence, and a life of self-actualization and well-being.

Well-being is defined as a state of happiness, contentment, lack of distress, overall good physical and mental health, and a healthy outlook with good quality of living. Maslow defined self-actualization as

> an episode in which powers of the person come together in an efficient and enjoyable way and in which he/she was

44 R.R. Llinas, 2201, *I of the Vortex: From Neurons to Self*, Cambridge: The MIT Press, pp. 21-23 and 127-28.

> more integrated and open to experience, more spontaneous and expressive, fully functioning, more creative, humorous, more ego-transcending, and more independent of his/her lower needs. He becomes more truly himself/herself more perfectly actualizing his/her potentialities, closer to the core of his/her being, more fully human.[45]

The self-actualized individual learns to overcome the four existential fears, the four = I's, namely Injury, Isolation, Insignificance and Inexistence. Yalom[46] wrote about the fears of one's non-existence or death, lack of external support and being on oneself with a need for free choice with responsibility, self-isolation from all creatures in the world, and a lack of meaning and hope in life. He suggested whole-hearted engagement in current life, creativity, self-actualization, and altruism as possible solutions to these ultimate existential fears.

Vedic seers have beautifully expressed the value of freedom from all fear and development of friendliness with others and Nature.

> अभयं मित्रादभयममित्रादभयं अभयं ज्ञातादभयं पुरो यः ।
> अभयं नक्तमभयं दिवा नः सर्वा आशा मम मित्रं भवन्तु ।।
>
> *abhayaṁ mitrādabhayamamitrādabhayaṁ abhayaṁ*
> *jñātādabhayaṁ puro yaḥ* ।
> *abhayaṁ naktamabhayaṁ divā naḥ sarvā āśā*
> *mama mitraṁ bhavantu* ।।
> — *Atharvaveda*, XIX.xv.6

> May we be fearless of the friend as well as the enemy, fearless of the known and the unknown, fearless during the day and the night. My hope is that all will be my friend.

The wisdom of self-integrity is beautifully expressed by

---

45 A.H. Maslow, 1968, *Toward a Psychology of Being*, 2nd edn., New York: Van Nostrand.

46 I.D. Yalom, 1980, *Existential Psychotherapy*, New York: Basic Books.

Erikson. It means acceptance of oneself and all aspects of one's life and integration of these elements into a stable pattern of living. It implies the experience of and adjustment to the trials and troubles of life as well as to its rewards and joys. Consequently, existence holds no fear; the ego has resigned itself to acceptance of life itself and to acceptance of the end of that life in death.

Integrity thus represents the fully developed personality in its most mature self-realization. Integrity is acceptance of one's own and only life cycle, and of the people who have become significant to it as something that had to be and that, by necessity, permitted of no substitutions. "The individual relinquishes the wish that important people in his/her life had been different and is able to love in a more meaningful way — one that reflects an acceptance of responsibility for one's own life."

The individual in possession of the virtue of *wisdom* and a sense of integrity has room to tolerate the proximity of death and achieve what Erikson termed a detached yet active concern. The style of integrity developed by his/her culture or civilization thus becomes the "patrimony of his/her soul. . . . In such final consolidation, death loses its sting."[47]

The Self as Integrated Information

Self is a complex, integrative process of a living organism. The living organism is a functioning, behaving unit of life. The living organism organizes, coordinates, and integrates information-energy within and around itself, spontaneously, subconsciously, and consciously. West[48] summarized the importance of neural integration:

How can diverse neuronal mechanisms provide coherent

---

47 D.S. Newton, 2005, "Erik H. Erikson", in B.J. Sadock and V.A. Sadock (eds.) *Kaplan & Sadock's Comprehensive Textbook of Psychiatry*, 8th edn., vol. 1., Philadelphia, Lippincott: Williams & Wilkins, p. 751.

48 J.B. West (ed.), 1992, *Best and Taylor's Physiological Basis of Medical Practice*, 12th edn., Baltimore: Williams & Wilkins.

> subjective experiences and volition? The answer is *integration*. Every living creature — beginning with the most primitive — is fully integrated; without integration, life is not possible. . . . What we are concerned with is not the search for a liaison between brain as a physical "thing" and some other entity, which we call "mind". We need to understand brain as a functioning tissue that generates feelings, moods, and self-awareness that constitute the basic ingredients of mind.

Tononi has recently proposed a theory of consciousness called "Integrated Information Theory", claiming that consciousness is integrated information. The quantity of consciousness is given by the amount of integrated information generated by a complex of elements, and the quality of conscious experience or "qualia" is given by the set of informational relationship within that complex.[49]

Since self-process is both convergent and divergent, it can function both as the source and the target of its behaviour and mentation. It is ever renewing since it is continuously generated as a result of spontaneous activation of cerebral cortex by the brainstem–reticulo–thalamic arousal system. The brainstem–cortical arousal system is crucial for the manifestation of a living, thinking self. Without such neurophysiologic arousal, consciousness, sense of self, attention, and volition are not possible.

The self process can be oriented in three complimentary modes of being-in-existence, namely egocentric, allocentric, and ecosystemic or existential mode. The first two interactive modes of reflective consciousness create the subject–object duality, whereas the existential, pre-reflective,

---

[49] G. Tononi and D. Balduzzi, 2009, "Toward a Theory of Consciousness", in S. Laureys and G. Tononi (eds.) *The Neurology of Consciousness: Cognitive Neuroscience and Neuropathology*, Burlington: Elsevier, pp. 1201-17.

consciousness is spontaneous, and non-dual. Such a mode can be intuitive and creative. The existential, conflict-free sphere of self-awareness remains the same throughout an individual's lifetime. The existential ecosystemic mode of being is consistent with the principles of "quantum neuroscience" and "neuro-ecology".

Homeokinetic Mind, Equanimity (Sthita-prajña), and Self-renewal

It is interesting to note that the modern homeokinetic physics is not only consistent with, but also provides a scientific explanation for the age-old introspective observations from the Vedāntic Indian literature. The ancient concept of mental equanimity (*sthita-prajña*) with self-mergence of mental activity can now be better understood in the light of homeokinetic physics. Homeokinetics is a logical extension of the commonly used term "Homoeostasis" in general Biology.

Homoeostasis is a well-established term in biology. "Homeo" means the same or similar and "stasis" means a steady state of equilibrium. Homoeostasis is defined as a tendency toward a relatively stable internal environment in organisms through interacting physiological processes and self-regulation.

Claude Bernard, a nineteenth-century French physiologist, first recognized the tendency toward stabilization of the internal environment of the animal body. In 1932, Walter B. Cannon, the American physiologist at Harvard University, used the term "homoeostasis" and published his well-known book, *The Wisdom of the Body*. Cannon described the fundamental adaptive process of homeostasis as the ceaseless balancing and rebalancing of physiological processes that maintain internal stability and restore a normal state when it has been disturbed.[50]

[50] C.P. Hickman Jr and L.S. Roberts, 1994, *Biology of Animals*, 6th edn., Dubuque, IA: Wm C Brown Communications, pp. 197-221.

Homoeostasis applies to many physiological variables at multiple levels of organization in both unicellular as well as multicellular organisms. Maintenance of physiological variables within a certain range of tolerance ensures optimal functioning of an organism. Some of the examples of homeostatic regulation in humans and animals include the maintenance of blood pressure, heart rate, blood oxygen, carbon dioxide, H^+-ion concentration or pH, water metabolism, osmotic pressure, body temperature, energy metabolism, neuro-endocrine regulation, and behavioural self-regulation, including our conscious decision-making. Homeostatic regulation applies to all levels of organization including molecular, cellular, tissue, organ system, individual organism, and even inter-individual social interactions.

Antonio Damasio recently proposed an interesting hypothesis about the possible relationship between subconscious or automatic homeostatic mechanisms and the evolution of adaptive conscious behaviour. He states:

> Homeostatic regulation, which includes emotion, requires periods of wakefulness (for energy gathering); periods of sleep (presumably for the restoration of depleted chemicals necessary for neuronal activity); attention (for proper interaction with the environment); and consciousness (so that high level of planning of responses concerned with individual organism can eventually take place).

Damasio recently summarized his position eloquently:

> Consciousness permits an extension of these automatic homeostatic mechanisms by allowing for flexibility and planning, important functions in complex and unpredictable environments. Conscious organisms know about their past and can make guesses about their future.

> They can implement this knowledge and manipulate it through planning, in an endeavour to approach that which is beneficial and avoid the harmful.[51]

Homoeostasis reminds me of the ideal meditative state of one's mind described in the *Bhagavad-Gītā* as a flame that remains unperturbed in a wind-free space: निर्वात दीप इव (*nirvāta dīpa iva*). The term "homeokinetics" was coined by Soodak and Iberall.[52] They consider the origin of life as a self-organizing process and propose ten principles to describe the physics for complex living systems.

> Self-organizing systems use cyclic processes to regulate their inner environments. These cyclic processes transport matter and energy and therefore in effect, import and export entropy. We call this thermodynamic cyclic conception of the life phase of complex systems *homeokinetics*, by extension of Claude Bernard and Walter Cannon's concept of homoeostasis. In the homeostatic view, internal mechanisms regulate the organism so as to return the organism to equilibrium. Homeokinetics emphasizes the role of kinetic processes, particularly in the form of ongoing periodic thermodynamic engine processes, in carrying out the regulating functions of the organism.
>
> A river represents a typical homeokinetic system. A river is not a fixed form, but is a process cycle. It is not a hydraulic element in a fixed channel, but the effect of rainfall in a hydrological cycle. The rainfall causes both water runoff and the appearance of a bedload of transportable material

---

51 A.R. Damasio and K. Meyer, 2009, "Consciousness: An Overview of the Phenomenon and its Possible Neural Basis", in S. Laureys and G. Tononi (eds.) *The Neurology of Consciousness: Cognitive Neuroscience and Neuropathology*, Burlington: Elsevier, pp. 3-14.

52 A.S. Iberall and H. Soodak, 1987, "A Physics for Complex Systems", in F.E. Yates, A. Garfinkel, D.O. Walter and G.B. Yates (eds.) *Self-Organizing Systems: The Emergence of Order*, New York: Plenum Press, pp. 499-520.

that erodes and maintains the form of the river.

The concept of a river as a homeokinetic system was expressed in one of my poems titled, *A River and a Lake:*[53]

A river is lake
that is in a hurry to reach the ocean;
A lake is a river
that pauses to look at the blue sky.
A river carves out of time
and knows the "time";
A lake reflects on eternity
and knows the "timeless".
A skilful mind can flow like a river
and stand still like a lake.
It depends on the terrain it faces
at the each moment.
It lives with spontaneity and joy,
Sometimes like a river,
and at times like a lake!

Another good example of a homeokinetic system is a school of fish swimming freely in an ocean. Each individual fish is an element or atomism of the system, but the school as a whole, behaves like a single holistic entity or an organism. Scott Kelso and David Engstrom suggest that the nature is experienced by humans as complementary pairs. The sense of self or agency of action emerges from the spontaneous self-organizing co-ordination tendencies. Meaningful information is the joint product of a co-ordinated system of parts and processes, which spans organism and environment. Co-ordination establishes meaning. The conscious awareness of self springs from the ground of spontaneous self-organized activity.

They described the ontogeny of our sense of self:

As a foetus moves spontaneously within the womb, it

---

53 V.D. Deshmukh, 2002, *A Poet's Walk: A Collection of Poems, Essays, and Photographs*, Florida: Jacksonville, p. 26.

> eventually discovers arms and legs that flex~extend, a mouth that opens~closes, a body that bends~twists. In other words, the neonate begins to make sense of itself as a living thing. "These movements belong to me! I am doing this!" From spontaneous self-organized behaviour emerges the self — "I am", "I do" — and from there a huge range of potentialities, "I can do". "I-ness" thus arises from spontaneity, and it is this "I" that may be said to direct action.[54]

Equanimity in Bhagavad-Gītā and Brahma-Sūtras

In the *Bhagavad-Gītā*, a person of self-realization with perfected wisdom (*prajñā*) is described as *sthita-prajña*. Such a person is described as one who has outgrown egoistic desires and delusional conflicts, who is self-content (*ātmatuṣṭa*), wise, and at peace with self (*sthitadhīḥ*). He is free of extreme negative or positive emotions, and resilient to changing circumstances. He has mastered the art of self-regulation (*saṁyama*) and he is free of episodic feelings of I-Me-Mine (*ahaṁ bhāva*). He realizes the true nature of self and reality. He looks at everyone with an eye of equanimity. Unperturbed by changing events, he enjoys spontaneous inner bliss (*ātma-prasāda*). He is compassionate (*karuṇa*) and cares for all living beings and Nature. He is ever-free (*mukta*) and a light to himself (*svayam jyotiḥ*).[55]

Brahma-Sūtras described the process of self-mergence (*svāpyayāt*) of the momentary ego (*kṣaṇika jīva*) into the enduring, knowing self (*śāśvata jñaḥ*). During activity, ego appears different from its inmost source, but both are the same just as light and the source of light are the same due to self-effulgence (प्रकाश आश्रयत्वात् तेजस्वात् *prakāśa āśrayatvāt tejasvāt*). It

---

54 J.A. Scott Kelso and D.A. Engstrom, 2006, *The Complementary Nature*, Cambridge, MA: The MIT Press, pp. 104-06.

55 Swami Madhavananda (Dr. Madhava Nagarkar), 2008, *Sat-Buddhiyoga*, Pune: Swaroopyoga Pratishthan.

describes the state of equanimity and stillness (*acalatvam*) after mergence of speech (*vāk*) into mind (*manaḥ*), mind into spontaneous breath energy (*prāṇa-śaktiḥ*), breath awareness into self-awareness (*ātmajñāna*), and self into the witnessing conscious energy (*cit-śaktiḥ*), which is its true nature (*tat ātmakatvam*). Such a person is truly free (*mukta*) and has no other ruler (*ananya-adhipatiḥ*).[56]

The main thesis of this section is that all of our mental processes, like conation, cognition, emotion, perception, attention, memory, intelligence, ego, and imagination, are homeokinetic and that they regularly go through periodic cycles of activity and rest. Whereas rest is a state of relative equilibrium or homeostasis, homeokinetic activity includes phases of emergence (*udaya*), existence (*vr̥ttiḥ*), and mergence (*laya*) of energy-information, of both conscious as well as subconscious experience.

The possibility of a spontaneous mergence of many of the transient mental activities into the spontaneous/intrinsic, matrix consciousness, which can still endure without any mental specificity, was clearly considered and discussed in a very recent chapter by Tononi and Laureys. The authors have successfully differentiated between intrinsic consciousness and almost all other mental functions like (i) sensory input-motor output, (ii) language, (iii) reflection/introspection, (iv) attention, (v) memory, (vi) sense of personal space, (vii) sense of body, and (viii) self/ego. They have convincingly argued that all of these mental functions can be dissociated in different circumstances from essential intrinsic consciousness with an existential sense of being. Their final conclusion is remarkably similar to the Vedāntic concept of spontaneous sound of peaceful silence (*anāhata nāda*), and the existential, witness consciousness (*sākṣī, ātman*). Here is their quote:

---

56 Swami Gambhirananda, 1965, *Brahma-Sutra Bhashya of Shankaracharya*, Calcutta: Advaita Ashrama, 1st edn., sixth impression 1996.

> Perhaps the states achieved through transcendental meditation techniques can approximate a loss of awareness of space, body, and self: experts report a feeling of silence, no bodily feelings, unboundedness (no space, and possibly no self), and no time. Even in such cases, though, the experience would still be happening to one entity — it would be "centred" if you wish, but not in the conventional spatial sense. This last sense of "self", which we may call "intrinsic" or "subjective" self — would be one and the same as the experience, not further dissociable.

A recent quote from J. Krishnamurti confirmed and eloquently described the factual experience of spontaneous sound of silence:

> Attention involves seeing and hearing. We hear not only with our ears but also we are sensitive to the tones, the voice, to the implication of words, to hear without interference, to capture instantly the depth of a sound. Sound plays an extraordinary part in our lives: the sound of thunder, a flute playing in the distance, the unheard sound of the universe; the sound of silence, the sound of one's own heart beating; the sound of a bird and the noise of a man walking on the pavement; the waterfall. The universe is filled with sound. This sound has its own silence; all living things are involved in this sound of silence. To be attentive is to hear this silence and move with it.
>
> — *Letters to the Schools*, vol. II, p. 30

Neuroscience of Meditation

Dhyāna-yoga is a Sanskrit word for meditation derived from the verb *dhyāyati*, which means to meditate, ponder, or be self-aware. By practice of *dhyāna*, one can achieve a state of intuitive, self-absorptive, spontaneous consciousness *per se* with serenity and presence of mind, which may lead to *samādhi* (enlightenment). Such a mental state is conducive to complete self-integration and understanding of the truth of one's being in reality and the reality in oneself.

The word *dhyāna* was transformed into *jhāna* in Pāli, *ch'an* in Chinese, and *zenna* or *zen* in Japanese language. The word *yoga* is derived from the verb *yuj*, which means to integrate or unite. Yoga is a way of self-integration. It is a system of discipline that facilitates the ultimate goal of an adaptive, creative, and enlightened living.

The Meditative Mind and Yoga

From ancient times, people have been fascinated by the wonder of their own mind and the experience of the surrounding universe. This pondering, about one's being-in-reality, may have been the beginning of a meditative mind, and the origin of various philosophies and religions.

The origin of Yoga is difficult to establish since it was preserved mainly by oral tradition. There are references to Yoga in ancient Sanskrit literature like the *Śvetāśvatara* and *Maitrī Upaniṣad*, and the *Bhagavad-Gītā*. Ghosh,[57] in his book *The Original Yoga*, included three books — *Śiva Saṁhitā*, *Gheraṇḍa Saṁhitā*, and *Pātañjala Yoga-Sūtras* — as the basis of Yoga.

Here are a few excerpts from the *Śvetāśvatara Upaniṣad*:

> Let the wise man restrain his mind vigilantly as (he would) a chariot with vicious horses. — Verse 9
>
> Even as a mirror stained by dust shines brightly when it has been cleaned, so the embodied one, when he has seen the real nature of the Self, becomes integrated, of fulfilled purpose and freed from sorrow. — Verse 14

A few excerpts from *Maitrī Upaniṣad*:

> That which is non-thought, which stands in the midst of

---

[57] S. Ghosh, 1999, *The Original Yoga*, New Delhi: Munshiram Manoharlal.

> thought, the unthinkable, the hidden, the highest, let a man merge his thought there. Then will this living being be (free) without any support (or dependence). — Verse 19
>
> . . . For by the serenity of thought, one destroys deeds, good and evil, with the serene self abiding in itself, he enjoys eternal happiness. — Verse 20

A few excerpts from the sixth chapter of the *Bhagavad-Gītā* on *dhyāna-yoga*:[58]

> Let a man lift his self by his own self; Let him not lower himself; One's self alone is one's own friend or foe.
>
> — Verse 5
>
> That in which thought ceases, stopped by the practice of disciplined attention, and in which, seeing himself through himself, one is content in oneself. — Verse 20

A few excerpts from *Śiva Saṁhitā*:

> Out of consciousness are born all the animate and inanimate objects (experiences) of this world. Hence, abandoning all these, one should abide in consciousness itself. — Verse 49
>
> . . . Indivisible by time and space, *ātman* in its natural state is always truly complete/whole. — Verse 54
>
> The *ātman*, however, is ever happy. — Verse 57
>
> . . . The *ātman* alone is without duality and beyond words.
>
> — Verse 61
>
> . . . All desires disappear only through knowledge (self-awareness). When the sense of plurality fades away, the light of Oneness dawns. — Verse 57

Excerpts from *Gheraṇḍa Saṁhitā*:

> Getting into (an absorptive) mental trance, let the mind be united with the self (*ātman*). When both (mind and self) merge

---

58 A.T. de Nicolas, 2004, *The Bhagvad Gita: The Ethics of Decision-Making*, Berwick, ME: Nicolas-Hays.

into the supreme soul (*paramātman*), *samādhi* ensues.
— Verse 16

. . . All living creatures on land and in the sky, all trees, bushes, creepers, plants, grass, all rivers, and mountains — all are seen as the one supreme *Brahman*, and one's own self is seen in all. — Verse 19

Yoga is a way of adaptive self-integration, being at peace with oneself and at home in the world. Self-integration results from constant self-awareness and letting go of irrelevant memories, imaginations, mental agitations, and distractions. Such a person functions with a truly undistracted and serene mind. When the mind is undirected and assumes its original, unmodified state, the self-experiences itself.

Yoga is a skilful quietening of a distracted and ruminative mind with a natural resolution of emotional conflicts, culminating in a truly peaceful and intuitive conscious state. Three essential steps of Yoga are —

1. *dhāraṇā:* conscious focusing of attention on an object, an activity, a thought or an ongoing experience;
2. *dhyāna:* continued, sustained dwelling of attention on the experiential field (outside and inside); and
3. *samādhi:* effortless, self-absorptive, and intuitive state of understanding and realization of one's true nature or being.

Here are a few excerpts from Patañjali's *Yoga-Sūtras*:

Yoga is calming the fluctuations of a (distracted and agitated) mind. — Verse 2

Then the self abides in his/her own (naturally serene) state.
— Verse 3

Other times, the self is associated with the agency (and ownership) of (ongoing) mental states. — Verse 4

> . . . Mental calming can be achieved by disengagement or letting go (of a ruminating mind) and constant self-awareness (presence, or mindfulness) — Verse 12

> The cause of selfishness or egoism is a lack of self-awareness. — Verse 75

> On becoming fully aware, the false association between self-regulatory activity and the essential self disappears, which leads to complete self-freedom. — Verse 76

> With vigilant attentiveness comes fitness of mind. — Verse 104

> . . . The capacity to change mental state manifests in three different ways: ability to stop ongoing mental activity, ability to focus attention on a specific event in space–time, and the ability to maintain attention on a given task or event leading to enhanced cognitive capacity. — Verse 122

> Being free from any self-centred motivation, instinctual pressure, and mental conflict, one can live naturally with spontaneity, presence, and natural wisdom. —Verse 195

> True freedom is freedom from oneself. When one is free, it is natural to be happy.

The Meditative Brain

We do not know exactly what happens in the human brain during meditation. Obviously, it is difficult to bring together individuals accomplished in meditation, during their enlightened state of mind, in a high-tech laboratory with functional Magnetic Resonance Imaging (MRI), Electro-encephalography (EEG), Magnetoencephalography (MEG), and Positron Emission Tomography (PET) scanning.

Lazar et al.[59] used functional MRI to identify the brain

[59] S.W. Lazar, G. Bush, R.L. Gollub, G.L. Fricchione, G. Khalsa and H. Benson, 2000, "Functional brain mapping of the relaxation response and meditation", *NeuroReport* **11**(7): 1581-85.

regions that are active during a simple form of meditation and relaxation response. A significant increase in signals was observed in the dorsolateral prefrontal and parietal cortices, hippocampus, temporal lobe, anterior cingulate, striatum, and pre- and post-central gyri during meditation. The results indicate that the practice of meditation activates neural structures involved in attention and control of the autonomic nervous system.

Newberg and Iversen[60] described two types of meditation: The first type is one in which the subjects simply attempt to clear all thoughts from their sphere of attention and to reach a subjective state, characterized by a sense of no space, no time, and no thought. This state is also fully integrated and unified, such that there is no sense of a self and other (*nirbīja dhyāna*). The second form of meditation involves focusing attention on a particular object, image, word, or phrase, which may lead to a subjective experience of absorption with the object of focus (*sabīja dhyāna*) (*fig.* 3.14).

Lutz et al. found some evidence that there is an increase in the fast gamma activity (25-42 Hz) and its ratio to slow oscillatory activity (4-13 Hz) on EEG during objectless meditation and non-referential compassion meditation. Gamma activity has been associated with attentional processes.

EEG and MEG have demonstrated spontaneous electromagnetic activity in brainstem and cortical areas. Spontaneous activity includes alpha rhythm, Mu rhythm, Tau rhythm, and the fast 40 Hz rhythm. The latter is also associated with conscious arousal, focal attention, and steady state auditory fields.[61]

---

60 A.B. Newberg and J. Iversen, 2003, "The Neural Basis of the Complex Mental Task of Meditation: Neurotransmitter and Neurochemical Considerations", *Med. Hypotheses*, **61**(2): 282-91.

61 R. Hari, 1999, "Magnetoencephalography as a Tool of Clinical Neurophysiology", in E. Niedermeyer and Lopes Da Silva F. (eds.) →

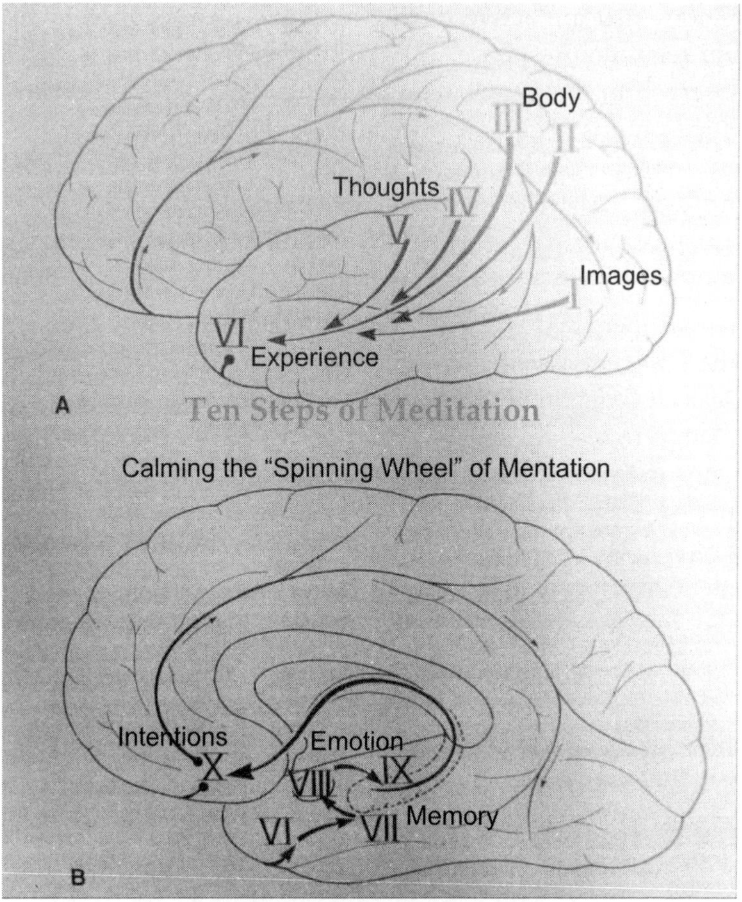

fig. 3.14: Ten steps of formal "sitting meditation with probable brain regions involved in the process". (Figure modified from C.R. Noback, N.L. Strominger, R.J. Demarest, and D.A. Ruggiero, 2005. *The Human Nervous System: Structure and Function*, 6th edn., p. 458)

Vedic Psychology: The Science of Wisdom

Most of the Vedic literature is devoted to self-culture (*ātmonnati*) and actualization of one's essential being with

→ *Electroencephalography: Basic Principles, Clinical Applications, and Related Fields*, 4th edn., Baltimore: Williams & Wilkins, pp. 1107-34.

innate wisdom. Wisdom is complete natural peace with one's being-in-reality. Vedānta variously described it as blissful conscious being (*sat-cit-ānanda*). What one experiences, from moment to moment, depends on one's perspective or "mind-set". As Hanumān expressed his feelings towards Lord Rāma:

> When I consider myself as the physical body (*deha-bhāva*), I am your servant/prisoner; when I consider myself as a living person (*jīva-bhāva*), I am an integral part of you, and when I consider myself as the essential being (*ātma-bhāva*), I am You.

The *ātma-bhāva* perspective is only possible in everyday life, after deep self-understanding and self-actualization. This wise stage in life is described in the Upaniṣads and the *Bhagavad-Gītā* as superconscious equanimity (*turīya, sthita-prajña,* ātma-prasāda, or *brāhmī-sthiti*).

There have been many examples of such enlightened individuals in human history. Therefore, such wisdom is potentially realizable by all of us. Understanding Vedānta, Yoga, and living mindfully with an undistracted presence can help us flower in virtue.[62] Yoga proposes the development of peacefulness, truthfulness, righteousness, fidelity, non-greed (*yama*), cleanliness, content, austerity, self-observation, and self-surrender (*niyama*).

Modern psychology and psychiatry also realize the value of wisdom as a virtue. Erikson conceives eight stages of self-development from birth till death: hope, will, purpose, competence, fidelity, love, care, and wisdom.[63] In his balance theory of wisdom Sternberg emphasized the importance of

---

62 Swami Tejomayananda and Swami Shantananda (eds.), 2008, *Living in the Present*, Langhorne: Chinmaya Publications.

63 B.J. Sadock and V.A. Sadock, 2007, *Kaplan & Sadock's Synopsis of Psychiatry: Behavioural Sciences/Clinical Psychiatry*, 10th edn., Wolters Kluwer / Lippincott Williams & Wilkins/Health, Philadelphia PA, pp. 207-11.

applying one's intelligence toward a common good through balancing of one's own, others', and institutional interests. The emotional intelligence comprises of emotional self-awareness, empathy, adaptation, stress-management, and happy optimism.[64] Wallace described the essentials for a meaningful life: genuine happiness, truth, virtue, and psychological flourishing by resolving conative, attentive, cognitive, and affective imbalances.[65]

Speeth wrote about "Therapeutic attention as the key to successful psychological counselling".[66] She emphasized the expansion of awareness to encompass all and everything, outside as well as inside of oneself. It is like witnessing self and environment together with impartial awareness (*ubhaya sākṣī*). It is an awareness of consciousness (meta-consciousness). Meditation and well-being go together naturally. It is a valuable skill to be discovered, well-learned, and lived whole-heartedly in everyday life.

Final Comments

Meditation is an art of efficient and adaptive management of neurobehavioural energy with full control over engagement and disengagement. In both of these conscious states, there is an experience of spontaneous, unitive being without a sense of ego, memory, or time. There is a natural sense of well-being with self-understanding, spontaneous joy, serenity, freedom, and self-fulfilment. It is where the ultimate pursuit of happiness and search for meaning in life come to fruition. One realizes the truth of one's being in harmony with nature

---

64 G. Matthews, M. Zeidner and R.D. Roberts, 2002, *Emotional Intelligence: Science & Myth*, Cambridge: The MIT Press.

65 B.A. Wallace, 2007, *Contemplative Science: Where Buddhism and Neuroscience Converge*, NY: Columbia University Press, pp. 2-11.

66 K.R. Speeth, 2003, "On Therapeutic Attention", in M. Brady (ed.) *The Wisdom of Listening*, Boston: Wisdom Publications, pp. 83-107.

and nature in oneself. Then each conscious moment becomes a dynamic process of discovery and learning of the ever-new reality. One's daily living becomes joyful (*ānandita*) and ever enlightening beyond all measure and description.

Attention can be voluntarily or involuntarily engaged or disengaged. An attentional act creates the duality of the attended–attendee, resulting in a specific conscious experience. Disengaging attention to the limit results in a serene attentiveness with holistic, non-specific (*nirguṇa*) intrinsic awareness. It may be expressed as the great nothingness (Buddha's *śūnyatā*), silence, or essential consciousness. Engaging attention to the limit, with a total participation in the present actuality, without any effort, sense of time, or self-consciousness, results in conscious totality (*pūrṇatā*). Between two consecutive attentional acts, there is inter-awareness, with no attended–attendee duality. This results in a refreshing "experience of no experience", which is analogous to waking up from a sound, dreamless sleep. This inter-awareness can lead to a state of thought-free super-consciousness (*turīya*).

Inter-awareness is like the gist consciousness, which can occur within 30 msec, whereas conscious attention takes about 300-500 msec. So, the gist awareness is 10-15 times faster than the conscious attentional process. My additional hypothesis is that both of these pre-attentive and attentive processes are initiated simultaneously and they run in parallel. However, they are completed at different times, namely 30 msec and 300-500 msec, due to shorter and longer neural pathways and the neural networks involved.

The pre-attentive processing occurs subcortically, whereas attentive processing needs complex and slow neo-cortical networks. Whereas conscious attentive processing is intermittent and "particulate", the pre-conscious, pre-attentive processing has "continuum" like properties. That is why the pre-conscious processing feels indivisible, "all-pervasive" and "eternal". The pre-consciousness and conscious attentiveness together make up the holistic superconsciousness (*pūrṇa, akhaṇḍa*).

Neurobehavioural energy (*prāṇa, cit,* or *acintya śaktiḥ*) and mentation can be in a potential or kinetic form. They are in kinetic form during common, interactive consciousness and dream sleep, whereas, during non-REM sleep and non-REM alertness (*turīya*), they are in a potential form. During non-REM sleep, the conscious arousal is low, whereas during non-REM alertness, the conscious arousal is high. That is why the *turīya* state has been described to have a sense of boundless energy, unlimited capacity, creativity, and an intense sense of well-being.

The moment one resumes directed attention, the subject–object duality returns, the "potential" neurobehavioural energy becomes "kinetic", and the egoistic mind takes on various conscious experiences. The ego continuously struggles to change the situation or itself, and therefore, suffers from mental pain and pleasure. There is a reciprocal relationship between the goal-directed, task-specific, dorsal (prefrontal–parieto–temporo–occipital) neural network and the resting medial (prefrontal–medial posterior cortical, precuneus, posterior cingulate gyrus, and retrosplenial cortex) "default mode neural network". When one network is active the other one becomes quiet, and vice versa.

Meditation, on the other hand, is the art of being alert and serene in the midst of any situation. It is the ultimate, adaptive reality–consciousness, continuously learning and renewing itself. Then the first and the second-person perspectives merge. *Ubhaya sākṣī* becomes *ātma-tuṣṭa*. The "witness consciousness" remains cheerful and blissful. It is the art of mastering both skills of parallel and sequential mental processing like the "River and Lake".

Ongoing meditation is the most essential state of one's being. One discovers and appreciates the wonders of this ever-new mode of living. However, the spontaneous "essence" remains the same. सारे क्षण सारखेच! (*sāre kṣaṇa sārakheca*). All moments are the same! What remains is the infinite, ineffable, essence, the actuality, the truth. Learning is endless. Life is good, when wisely lived.

आत्मनि एवं आत्मना तुष्टः।
ātmani evaṁ ātmanā tuṣṭaḥ।
One becomes naturally blissful within oneself.

तरति शोकं आत्मविद्।
tarati śokaṁ ātmavid।
Knower of Self overcomes all grief.

ब्रह्मार्पणं ब्रह्म हविर्ब्रह्माग्नौ ब्रह्मणा हुतम्।
ब्रह्मैव तेन गन्तव्यं ब्रह्मकर्मसमाधिना।।
brahmārpaṇaṁ brahma havir brahmāgnau brahmaṇā hutam।
brahmaiva tena gantavyaṁ brahma-karma-samādhinā।।
— *Bhagavad-Gītā* IV.24

The existential process is *Brahman*. The matter (clarified butter or fuel) offered to the fire is *Brahman*. The agent, the person making the offering is *Brahman*. The sacred fire, the transforming energy is *Brahman*. One who realizes this holistic process of existence, reaches (the one and only) *Brahman* alone.

शान्तोयम् आत्मा।
śāntoyam ātmā।
The self lives blissfully, silently!

ब्रह्मविद् ब्रह्मैव भवति।
brahmavid brahmamaiva bhavati।
Knower of *Brahman* becomes *Brahman* itself.
One lives embedded in LIFE itself.

(*Brahman*: The ineffable, infinite, absolute blissful, conscious being-in-existence/energy.)

Acknowledgement

I wish to thank Dr Sunanda V. Deshmukh, MD, for her careful review of the manuscript and making some valuable suggestions.

Appendix

Additional Vedic Quotations

एकं ज्योतिः बहुधा विभाति।
ekaṁ jyotiḥ bahudhā vibhāti ।
One flame/energy shines in many ways.

असतो मा सद् गमय। तमसो मा ज्योतिर्गमय।। मृत्योः मा अमृतं गमय।।
asato mā sad gamaya । *tamaso mā jyotirgamaya* ।। *mr̥tyoḥ mā amr̥taṁ gamaya* ।।
Guide me from unreal to real. Guide me from darkness of ignorance to the light of knowledge.

Guide me from fear of death to a blissful life.

देहो देवालयः प्रोक्तः स जीवः केवलः शिव।
त्यजेत् अज्ञान निर्माल्यं सोऽहं भावेन पूजयेत्।।
deho devālayaḥ proktaḥ sa jīvaḥ kevalaḥ śiva ।
tyajet ajñāna nirmālyaṁ so 'haṁ bhāvena pūjayet ।।

This body is the temple. The living being is sacred. Let us discard the old flowers of self-ignorance, and worship self as the Divine.

अपाणि पादोः अहम् अचिन्त्य शक्तिः। पश्यामि अचक्षुः स श्रुणोमि अकर्णः।।
अहम् विजानामि विविक्त रूपो।। न च अस्ति वेत्ता मम चित् सदाहम्।।
apāṇi pādoḥ aham acintya śaktiḥ । *paśyāmi acakṣuḥ sa śruṇomi akarṇaḥ* ।।

aham vijānāmi vivikta rūpo || *na ca asti vettā mama cit sadāham* ||

I am that incomprehensible energy that acts without hands or feet, that can see without eyes and hear without ears. I can comprehend various forms, but, no one knows me as the timeless awareness.

तदेजति तन्नैजति तद्दूरे तद् वन्तिके।
तदन्तरस्य सर्वस्य तदु सर्वस्यास्य बाह्यतः।।
tadejati tannaijati taddūre tad vantike |
tadantarasya sarvasya tadu sarvasyāsya bāhyataḥ ||
It moves. It moves not. It is far. It is so close.
It is within all. It is beyond all.

स्वल्पापि दीप कणिका बहुलं नाशयेत् तमः।
स्वल्पापि बोधो निबिडं बहुलंनाशयेत्तथा।।
svalpāpi dīpa kaṇikā bahulaṁ nāśayet tamaḥ |
svalpāpi bodho nibiḍaṁ bahulaṁnāśayettathā ||
Even a brief spark of light can eliminate dense darkness. Similarly, even a brief insight (understanding) can eliminate deep ignorance.

अन्तःपूर्णो बहिःपूर्णः पूर्णकुम्भ इवार्णवे।
अन्तःशून्यो बहिःशून्यः शून्यकुम्भ इवाम्बरे।।
antaḥpūrṇo bahiḥpūrṇaḥ pūrṇakumbha ivārṇave |
antaḥśūnyo bahiḥśūnyaḥ śūnyakumbha ivāmbare ||

Submerged in an ocean, a pot is full within and without. Being in the sky, a pot remains empty within and without.

पूर्णमदः पूर्णमिदं पूर्णात् पूर्णं उदच्यते।
पूर्णस्य पूर्णमादाय पूर्णमेवावशिष्यते।।
pūrṇamadaḥ pūrṇamidaṁ pūrṇāt pūrṇam udacyate |
pūrṇasya pūrṇamādāya pūrṇamevāvaśiṣyate ||
This is whole. That is whole. From the whole emerges whole. Even after the emergence of whole, the original (Absolute) remains whole.

भूः भुवः स्वः तत् सवितुर्वरेण्यम्।
भर्गो देवस्य धीमहि धियो यो नः प्रचोदयात्।।
bhūḥ bhuvaḥ svaḥ tat saviturvareṇyam ।
bhargo devasya dhīmahi dhiyo yo naḥ pracodayāt ।।

We meditate on that supreme effulgence of the radiant Being within, the indwelling Guide within all beings. May it inspire our intellect.

श्लोकार्धेन प्रवक्षामि यदुक्तं ग्रन्थ सहस्त्रभिः।
ब्रह्म सत्यम् जगन् मिथ्या जीवो ब्रह्मैव नापरः।।
ślokārdhena pravakṣāmi yaduktaṁ grantha sahasrabhiḥ ।
brahma satyam jagan mithyā jīvo brahmaiva nāparaḥ ।।

I will declare the essential Truth in half a stanza, which has been elaborated, in thousands of books: The Absolute is the Truth. The experienced world is (transient) appearance. An individual is none other than the Absolute.

नाना उपाधीवशाद एव जातीवर्णाश्रमादयः।
आत्मनि आरोपितास्तोये रसवर्णादिभेदवत्।।
nānā upādhīvaśāda eva jātīvarṇāśramādayaḥ ।
ātmani āropitāstoye rasavarṇādibhedavat ।।
— *Ādi Śaṅkarācārya Ātmabodha*, 11

Only on account of diverse superimpositions like caste, colour, stage in life, etc. does the *ātman* feel different, just as the pure water that is tainted by different colours, flavours, etc.

न विज्ञाते: विज्ञातारं विजानीयाः।
na vijñāteḥ vijñātāraṁ vijānīyāḥ ।
The knower of the knower is not knowable.

चिन्मात्रोहं सदाशिवः।
cinmātrohaṁ sadāśivaḥ ।
I am the absolute awareness, the absolute goodness.

अत्र अयं पुरुषः स्वयंज्योतिः भवति।

atra ayaṁ puruṣaḥ svayaṁjyotiḥ bhavati ।

In this (Superconscious) state, the self becomes Self-Illumined.

(It becomes Self-effulgent, a light to itself.)

निर्वात दीप इव।

nirvāta dīpa iva ।

Tranquil awareness, like the unperturbed flame of a lamp.

आत्मनि संविशति आत्मनात् आत्मानम्।

ātmani saṁviśati ātmanāt ātmānam

The self submerges by itself into Self.

A Quotation from J. Krishnamurti

Meditation is to find out whether the brain, with all its activities, all its experiences, can be absolutely quiet. Not forced, because the moment you force, there is duality. The entity that says, "I would like to have marvellous experiences, therefore I must force my brain to be quiet" will never do it. But if you begin to inquire, observe, listen to all movements of thought, its conditioning, its pursuits, its fears, its pleasures, watch how the brain operates, then you will see that the brain becomes extraordinarily quiet; that quietness is not sleep but it is tremendously active and therefore quiet. A big dynamo that is working perfectly hardly makes a sound; it is only when there is friction that there is noise.[67]

---

[67] J. Krishnamurti, 1979, *Meditations*, San Francisco: Harper & Row Publishers, p. 4.

4

The Mystery of Time[1]
Converging Thought Currents from Vedānta, Physics, and Neurosciences

Raju Chidambaram and *Shobha Chidambaram*

Abstract

The cosmological theory of Vedānta and the relativity theory of modern physics both suggest a timeless universe in which there are no past–present–future distinctions. However, it is common experience that sentient creatures (*jīva*s) living in the universe do perceive time. A resolution of this inconsistency between the macrocosm and microcosm is possible using concepts in Vedānta as well as some of the emerging views in modern neurosciences. The resolution presents a picture of the cosmos that departs radically from common perceptions about the nature of our material existence in the universe, but that picture is nevertheless consistent with Vedāntic teachings about our spiritual nature.

Introduction

A REVOLUTION in scientific thought of profound significance has been unfolding since the early twentieth century when physicists

---

[1] Much of the material in this chapter is taken from the book *Law of Love and the Mathematics of Spirituality*, by Raju Chidambaram (2011).

began probing deeper into the nature of matter, energy and the cosmos. Earlier scientific view of a solid, absolute, deterministic universe which exists regardless of the observer, is being replaced by a fuzzy, relativistic, unpredictable world whose characteristics, if not the very existence, depend on the observer. The new scientific views bear a remarkable resemblance to the ancient Eastern philosophies as has been commented upon extensively by many great scientists and philosophers (Capra 1975). Vedānta, which is one of the most developed of these Eastern philosophical systems, in particular, has been often cited for its relevance to the modern theories of quantum mechanics and relativity (Swami Jitatmananda 1986; Goswami 1993). Exploiting the synergy between physics and Vedānta, Goswami has also shown how the principles of quantum mechanics can be used to explain Vedāntic concepts such as soul, reincarnation, telepathy, and liberation (Goswami 2001). A primary source of the synergy is due to the fact that the central role of a sentient observer in quantum systems parallels the overarching role of Consciousness in Vedānta.

The concept of time is emerging as yet another key aspect of modern science which ties in well with Vedānta. Einstein's Relativity Theory totally destroyed the long-held views of time as flowing from a dead-and-gone past to the present "now" to a yet-to-be-born future. Though physicists are still debating the true nature of time, it is now accepted by most, if not all, that this old view of time is flawed. Universe, they believe, is essentially timeless with no past–present–future distinctions (Callender 2010). Vedānta also holds that *Brahman* (the ultimate reality of which the universe is an apparent manifestation) is changeless and timeless. Time and all that which change with time are an illusion (*māyā*). Thus, while both modern science and Vedānta assert timelessness, it opens up an obvious question: How does a *jīva* experience time in a world that in truth is timeless?

This chapter is an attempt to answer this mystery behind time. We will now examine the various concepts of time starting with classical scientific view.

The Classical View of the Universe

Until Einstein's arrival on the scene, time was viewed by most Western philosophers and scientists as the firm and solid stage on which the grand drama of the universe was enacted. This classical view may be represented graphically as in *fig*. 4.1.

The state of the universe changes from moment to moment under the pressure of natural forces acting on the various objects and beings of the universe. The present state of the world, together with the forces currently acting on it, is the cause determining the state of the world at the next moment. Whereas physicists generally consider only material objects and their physical states in describing the universe at any given moment, we also include in our description of the world sentient beings (*jīvas*) and their mental states (of emotions and thoughts). Following a notation made popular by Swami Chinmayananda, the world of objects–emotions–thoughts is denoted by O-E-T (Swami Chinmayananda 1978). In *fig*. 4.1, time, *t*, is the substratum on which the changes in O-E-T take place.

Twentieth-century physics as well as Vedānta find this classical view (sometimes called the "Newtonian" or "clock work" universe) deficient and incomplete. We will see in a later section how the twentieth-century physics differs from classical physics, but will first present the Vedāntic concepts regarding time in the next four sections.

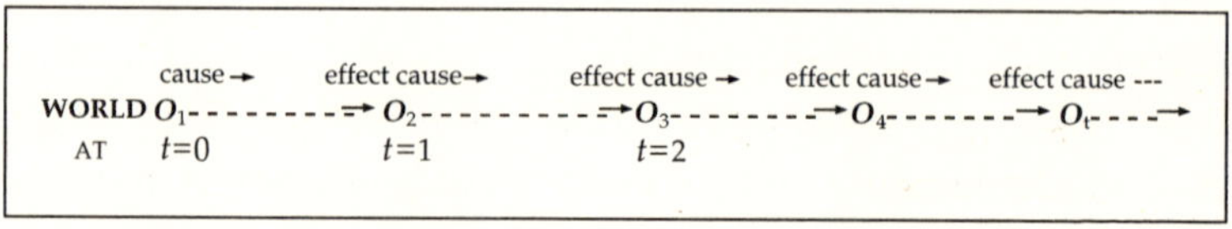

fig. 4.1: O_t, the world of objects-emotions-thoughts (O-E-T) at time t = 0, 1, 2, . . .

THE VEDĀNTIC MODEL

Two Problems with the Classical View

There are primarily two problems that the Vedāntins will have with the classical view. The first problem is with the suggestion in *fig*. 4.1, that the evolution of the world is entirely determined by the laws of nature. There is no room here for a "higher will" (*Īśvara*) to influence the affairs of the world. But, in Vedānta, as in most religions, the world is considered to be created and controlled by a superior will. The second problem is that the model does not account adequately for the fact that the world not only exists, but its existence is experienced. In fact, we know its existence only through experience. The changing world portrayed in *fig*. 4.1, is known and experienced by *jīvas*, that is, conscious beings such as humans. Now, according to the Vedāntic principles, an object known and the subject that knows it cannot be one and the same. Therefore, the subject that knows the world cannot be totally part of the world. It must be outside the one-dimensional line in *fig*. 4.1. This leads us to propose a two-dimensional model shown in *fig*. 4.2.

The one-dimensional line which was the substratum for the changing world in *fig*. 4.1 is now replaced by a two-

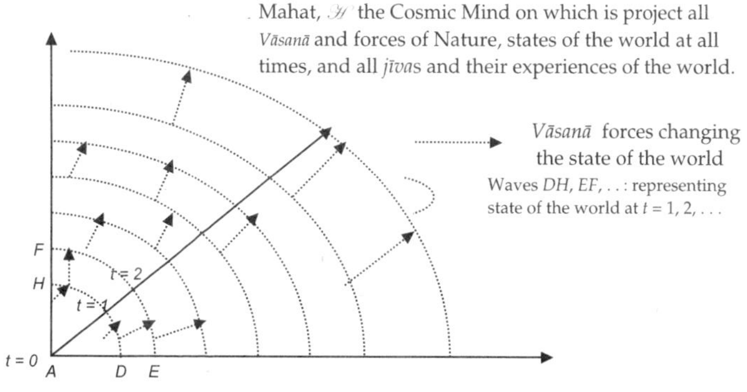

fig. 4.2: The two-dimensional space of experienced world and experiencing *jīvas*

dimensional space $\mathscr{M}$, bound by two axes at right-angle. For reasons to be provided in section 4 below, the space $\mathscr{M}$ is identified with the Vedāntic concept of cosmic mind (*mahat*) in which exist the entire created cosmos of things, beings and their experiences over time. The changing world, previously resembling beads strung on the timeline, now appears as concentric "waves" radiating from the origin *A*. For example, the wave *HD* in *fig*. 4.2 with the label $t = 1$, indicates the state of the world at time 1, while the wave *FE* represents the world at the "next" moment, namely $t = 2$.

What cause the world to go from one state at $t = 1$ to another state at $t = 2$ are the forces affecting both the gross physical matter and the subtle "mind-intellect" of sentient beings. The forces affecting physical matter, or the "forces of nature", follow the laws known to science. The forces affecting mind-intellect, called *vāsanā*s (tendencies) in Vedānta, are less understood, but equally real. The physical and "mental" forces together are the agents of change in O-E-T and the term *vāsanā*s can be applied for both. For example, under the pressure of the *vāsanā*s in effect at $t = 1$, the world moves to a different state, indicated by the wave *EF*, at $t = 2$. To use a more suggestive language, we may say that the winds of *vāsanā*s blowing at time $t = 1$ change the wave *HD* at time 1 into the wave *FE* at time 2. The arrows in *fig*. 4.2 show the *vāsanā* pressure behind the waves. This picture of the cosmos echoes the following beautiful simile employed by Ādi Śaṅkara to describe the perception of a Realized Being (Swami Chinmayananda 1978):

> In me, the ocean of unbroken Bliss, endless waves of the universe are created and destroyed by the play of the storm of *māyā*.[2]

---

[2] *Māyā* is a basic concept in Vedānta signifying, among other things, the creative force using which *Īśvara* projects the world on the cosmic mind. Everything in creation, including *vāsanā*s, is an effect of *māyā*.

Several points need be made to explain what the "waves" and the field in which they arise are. Each wave, *as a whole*, represents the state of the O-E-T at a particular time: for example the arc *DH* as a whole is the state of the world at $t = 1$. The phrase "as a whole" is used to imply that segments or points on the arc do not have any significance; in particular, they do not represent particular parts or elements of the world. Secondly, the state of the world includes not only an objective description of the physical world at a given time, but also a subjective description of that world from the point of view of each *jīva*. The physical world being the same, it is seen by different *jīva*s differently. What is a happy world to one *jīva* may be at the same time a sad one to another. This fact is conceptually accommodated in the model since the state description of the world includes the physical state as well as the emotions and thoughts specific to every *jīva*.

Another point to note is about the role of the Creator or God (*Īśvara*) in determining the state of the world at any time. As mentioned earlier the *vāsanā*s, the forces of nature (*prakṛti*), are behind the ever changing state of the world. However, the *vāsanā*s do not totally determine the changes. This is also the conclusion of modern science. *Vāsanā*s only suggest the different possible changes that could happen and their relative likelihood. Only one of these possibilities is actually realized and it is as willed by *Īśvara*. In *fig*. 4.2, *FE*, the world at $t = 2$, arises due to the *vāsanā*s in operation at $t = 1$ on *DH*. Knowing the state of the world *DH* and the *vāsanā*s in operation, we can at the best project the different possible states of the world at $t = 2$. *FE*, the actual state at $t = 2$, is chosen from the many possibilities at *Īśvara*'s will. Thus, the evolution of the cosmos, shown as the expanding field of waves in the space $\mathscr{M}$, is the combined result of *prakṛti*'s forces and *Īśvara*'s will.[3]

---

[3] The waves *DH, FE*, etc. shown in *fig*. 4.2 are the actual states of the world that the *jīva*s experience. Alternatively, we can consider each →

The Experiencer

The changing world, represented by the sequence of waves, is what is experienced by the sentient *jīvas* of the cosmos. That is to say, the "experienced" is shown by the waves at $t = 0$, $t = 1$, $t = 2$, etc. Thus the parameter t is also a proxy for the world experienced at time t. Those familiar with the polar co-ordinate system for two-dimensional spaces will recognize that t is the "radius vector" with A as the origin. Using polar co-ordinates, any point P in the space $\mathcal{M}$ can be designated by specifying the wave t on which it is located and the angle, θ, the line AP makes to the "initial line" (or "horizontal axis") AE. See *fig*. 4.3.

Since t refers to the experienced, one can expect the other parameter θ to relate to the "experiencers", or *jīvas*. This indeed is the case, but we need to be clear about the sense in which this is so. The parameter θ does not refer to any physical,

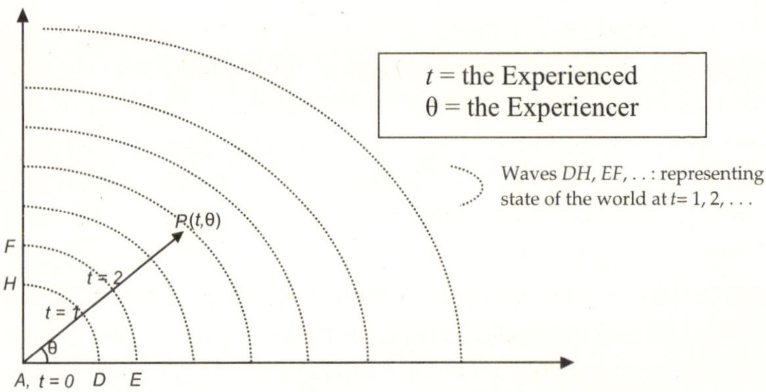

fig. 4.3: Polar co-ordinate designation of $\mathcal{M}$, the field of experiences

→ wave to represent *all* possible states the world can be at that time, and not just the actual state as decided by *Īśvara*. The field M is now the field of all possible experiences from the beginning of the creation to its end. This leads to a picture similar to the multiverse theory proposed by some quantum physicists.

mental, or intellectual states of the *jīva* since, by our definition, these are parts of the experienced O-E-T. Instead it refers to the *jīva*'s spiritual detachment, a measure of the spiritual maturity of the *jīva*.[4] As a *jīva* progresses spiritually, its spiritual detachment, θ, increases. In our model, θ, measured in radians, can take values from 0 to $\pi/2$. The horizontal and vertical axes in *fig*. 4.3 correspond respectively to the lower and upper limits of θ. A *jīva* at the lower limit of $\theta = 0$, represents a being with zero detachment; in other words it is a *jīva* which is fully attached to the world, which is ignorant of the Self, and suffers the dualities of pain and pleasure, etc. This *jīva* is in the "indulging mode" as it compulsively "eats" or savours whatever sweet or bitter fruits the world offers. In contrast, at the upper limit, $\theta = \pi/2$, is a *jīva* fully detached from the world, living in full knowledge of self, remaining in bliss while merely witnessing the changing world. This characterization of *jīva*s with differing spiritual detachment is in keeping with the basic teachings of Vedānta. *Muṇḍaka Upaniṣad*, for example, uses the simile of two friendly birds perched on the same tree (of life), one eating the fruits of the tree while the other sits merely looking around (Swami Chinmayananda 1988).

Our spiritual personality has many facets. One may therefore question if the entire spiritual nature of a *jīva* can be reduced to a single parameter, θ, denoting its spiritual detachment. A mathematical justification for this simplification has been provided in the work by one of these authors where it is shown that spiritual detachment, θ, of a *jīva* at a given

---

4 *Vairāgya* (spiritual detachment), of a *jīva* is the result of the wisdom to see (a) the ephemeral nature of the material world, and at the same time (b) the eternal, blissful nature of own true self. Such a *jīva* does not identify with the happenings of the world but remains detached from it even while fully transacting with the world. Spiritual detachment does not therefore imply disengagement from life.

time can be related to the mean value of a function representing its *vāsanās* at that time (Chidambaram 2008). In this sense, spiritual detachment, θ, is a convenient summary of the entire spectrum of *vāsanās* (ranging from base material instincts to the highest spiritual values) energizing the *jīva* at the moment. Based on the teachings of *Yoga Vāsiṣṭha*, Swami Tejomayananda also has noted that "one of the measuring rods of spiritual progress is our degree of spiritual detachment" (Swami Tejomayananda, 1998).

The $\mathscr{M}$-Space

The space defined by the coordinates (t,θ) $\{0 \leq t < \infty$ and $0 \leq \theta \leq \pi/2\}$ is the $\mathscr{M}$-space, and is the first quadrant of the two-dimensional space. $P(t,\theta)$ represents the point in the $\mathscr{M}$-space where a *jīva* with detachment θ, encounters the world at time t. The result of an encounter of an experiencer with the experienced is an *experience*. The points in space $\mathscr{M}$ are thus points of experience and the space $\mathscr{M}$ as a whole is the space, or field, of all experiences of all *jīvas* from time 0 to infinity. In other words, the $\mathscr{M}$-space represents the entire creation and its evolution over time. In Vedānta, creation is considered to be a projection on the *mahat*. On this basis, the $\mathscr{M}$-space is identified with the *mahat*. The *mahat* itself exists in *Brahman*, which is the ultimate "ground of being" for everything including *Īśvara, māyā,* the created world, the *jīvas* and their experiences.

Coexistence of Past, Present, and Future in the $\mathscr{M}$-Space

Creatures are in creation. If the two-dimensional (t, θ) plane in *figs*. 4.2 and 4.3 is compared to a wall in a room, creatures are like flies on the wall. In contrast, Creator is not in creation and is like someone in the room looking at that wall. Therefore the view of creation shown in *figs*. 4.2 and 4.3 is not available to a creature bound to the wall, but only to the Creator. Time and space are also part of creation and hence they are in the *mahat*. On the other hand, the Creator and the *mahat* on which the creation is projected are not in time or space. They are timeless entities and concepts such as past, present, and future, do not apply to them. They are also changeless, since change implies time. Such being the case, from the point of view of the creator it cannot be that the wave *DH* at time = 1 arises first and then disappears before the wave *EF* at time = 2 is born. If this were so, the Creator, who sees one world now and a different one next moment, will be also in time domain. Instead, from the Creator's point of view all the waves must exist together in the $\mathcal{M}$-space. That is to say, the "past, present, and future" worlds coexist along with all the experiences they have in store for all the *jīva*s. It is not that the past "was", the present "is" or the future "will be", but in fact the past, present and future all "is". There is only "is" ness in the higher realms of reality associated with *Īśvara* and *mahat*. This is the conclusion to be drawn from the timelessness of *Īśvara*.

The waves in *fig*. 4.3, have time tags attached to them, such as t = 1, t = 2, etc. Since there is really no time concept associated with the $\mathcal{M}$-space, the tags do not refer to time as we *jīva*s understand and experience it. We will refer to t as "cosmic time" to differentiate it from ordinary time, but should understand that, from the highest perspective, the tags refer to sequential ordering rather than time.

The $\mathcal{M}$-space of *fig*. 4.3, with wave upon wave spreading outwards from the origin *A*, resembles a modern-day CD disc with the entire story of the cosmos recorded on its tracks. This turns out to be a thought shared by other philosophers and mystics, one example being the concept of "ākāśic records" espoused by Edgar Cayce (Todeschi 1998). But of immediate interest to us is the similarity of this Vedāntic idea to Einstein's Relativity Theory.

VIEWS OF MODERN-DAY PHYSICS

The "Block Universe" Model of Relativity Physics

Relativity physics also views the universe in terms very similar to those of Vedānta. Einstein proposed, and later observations confirmed, that time and space are not independent of each other but form one integral whole, namely the "space–time continuum". Each point in this continuum refers to potential or actual "event". *Fig*. 4.4, adapted from Brian Greene's popular book *The Fabric of Cosmos*, shows the universe in space–time co-ordinates as the open rectangular box *LMNOPQ* (Greene 2004). Resembling a loaf of bread, this space–time representation is sometimes called as the "Block Universe". Time progresses from left to right along the x-axis. The cross-sections of the block labelled as A_1, A_2, and A_3 (which are like three slices of the loaf of bread) show the universe at three different points in time for some observer *A*. e_1, e_2, and e_3 are three points in the time–space continuum of which e_1 lies in the time-slice A_1 while e_2 and e_3 both are on slice A_2. This means that the observer *A* sees event e_1 first and sometime later sees e_2 and e_3 simultaneously. However, another observer *B*, who is in uniform motion relative to *A*, will in effect slice the space–time block at a different angle. Therefore, for this observer, both e_1 and e_2 may be in the same time-slice B_1, and e_3 on a

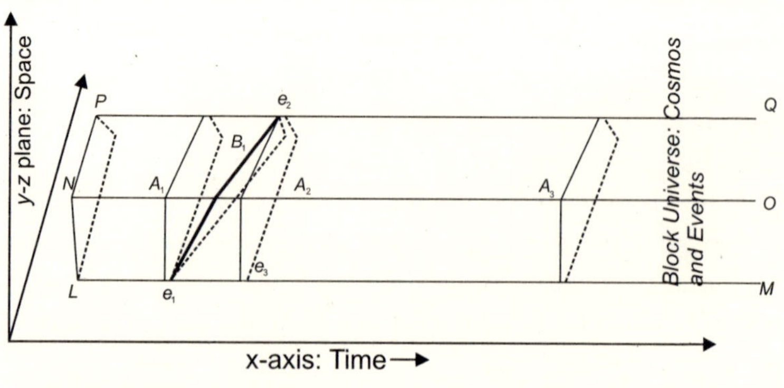

fig. 4.4: The Block Universe and space–time continuum

subsequent slice B_2 (not shown in the *fig.* 4.4). Thus *B* will see e_1 and e_2 happening at the same time, and e_3 by itself later.

This being the case, we cannot assert that e_2 happens only simultaneous with e_1 or only simultaneous with e_3. Both statements are correct from the respective observer's point of view. Therefore, we infer that the event e_2 exists not as part of any particular time-slice but as part of the block as a whole. The larger implication of this is a basic result of the Special Relativity Theory: The space–time continuum exists as a whole, along with all the events like e_1, e_2, and e_3, and not as individual time-slices. In *fig.* 4.4, the loaf of bread exists as a whole with no special significance attached to the different "slices" one may make of it. All events, past, present, and future, coexist in the Block Universe though observers *in* the universe see them sequentially. Past, present, and future have no meaning as seen from outside the universe, in this modern scientific conception of the cosmos.

These revelations of Relativity Theory are contrary to our common sense notions of time and space, but most thinkers have come to accept it as the truth in view of experimental evidence supporting the theory. As Einstein himself once observed, "To us convinced physicists, the distinctions between past, present, and future are illusory, however persistent". The famous astronomer Sir Arthur Eddington concurred: "Events do not happen: they are just there, and we come across them . . . (as) the observer on his voyage of exploration." Louise de Broglie, another famous physicist, stated the same view thus:

> Each observer, as time passes, discovers, so to speak, new slices of space–time which appear to him as successive aspects of the material world, though in reality the ensemble of events constituting space–time exist prior to his

> knowledge of them . . . the aggregate of past, present and future phenomena are in some sense given a priori.[5]

The "future" is not necessarily known a priori to any observer who is part of the universe, but only if someone has the view, as in *fig.* 4.4, of the universe from *outside* the space–time continuum. Observers in the universe do not have the timeless view possible from outside and thus are subject to time. Time is an artefact (or illusion, in Einstein's words) experienced by observers moving through space–time continuum.

Clearly, this space–time continuum picture of the modern-day physics shares many similarities with the $\mathscr{M}$-space, or *mahat*, of Vedānta. However, there is a major difference between the two. In the Vedāntic model, the emphasis is on the world as experienced by the *jīvas*. Thus we find one of the two polar co-ordinates in the Vedāntic model specifically designated for the experiencer. In physics, on the other hand, the observer is more of an afterthought, brought into the picture rather reluctantly only when needed to explain better the physical world. The space–time continuum description treats the observer more like a physical sensing device and less as a conscious entity that may have presence in dimensions other than the space–time continuum. Neuroscientists, who must deal with the world perceived by the sensory organs and the brain, often find themselves positioned somewhere in between the physicist and the Vedāntin. We will see later how at least one neuroscientist has tried to stride the divide. Before we do that, however, it is instructive to explore further the similarities between the Vedāntic and relativity physics models.

The Minkowski Metric and the Distance Measure in $\mathscr{M}$-Space

The correspondence between the Vedāntic concepts and

---

[5] Both quotes are as given in an article by J. Smythies (2003). Einstein's quote is from Brian Greene's book *The Fabric of Cosmos*.

Relativity Theory can be delineated more sharply using the so-called Minkowski's Metric for time–space continuum. To start the discussions, consider *fig.* 4.5 showing two points of encounter for a *jīva* with detachment θ. In this figure, we choose to represent the $\mathscr{M}$-space in Cartesian co-ordinates (τ, G) where $\tau = t.\cos\theta$ and $G = t.\sin\theta$. These co-ordinates can be assigned a specific spiritual meaning based on the Vedāntic theory, as will be seen shortly.

Let *AB* and *CD* be the two consecutive waves and *PQ* the vector at angle θ intersecting the two time arcs at *P* and *Q*, respectively. *P* and *Q* are the *jīva*'s two successive encounters, or experiences, with the world. Let δs denote the distance *PQ* between *P* and *Q*. δs denotes the "duration" of the experience from the *jīva*'s encounter with the world at *P*. The elemental distance δs in $\mathscr{M}$-Space is not time, but something more like a measure of the space–time. Let *PR*, denoted by $\delta\tau$, and *RQ* denoted by δG be the projections of *PQ* on the horizontal and vertical axes, respectively. From the geometry, it may be noted that $PR = \delta s.\cos\theta$, $RQ = \delta s.\sin\theta$, and $\delta\tau^2 + \delta G^2 = \delta s^2$. The fact that $\delta\tau$ and δG depend on θ makes it

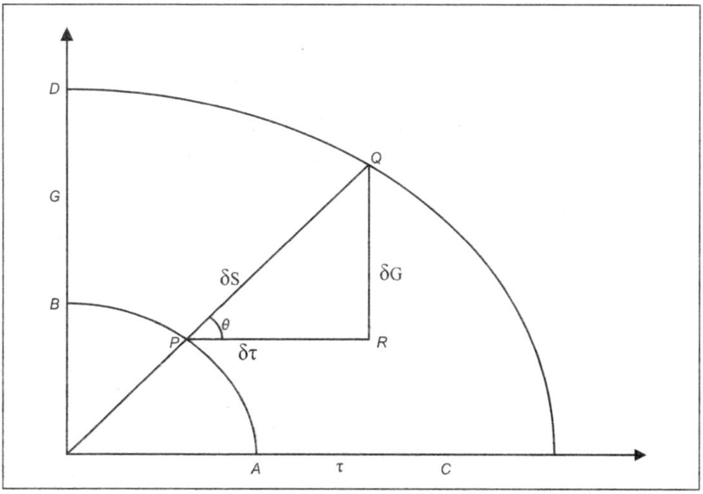

fig. 4.5 $\mathscr{M}$-space in Cartesian co-ordinates

possible to provide a spiritual interpretation of these variables based on Vedānta.

If a *jīva* is fully attached to the world, its spiritual detachment θ is zero and hence $PR = \delta\tau = \delta s$, whereas $\delta G = 0$. That is, the projection of PQ is wholly on the horizontal axis, which is the axis of zero detachment or total worldliness. Ignorant of self and totally lacking in detachment, such a *jīva* is fully attached to the experience from its encounter with the world at P. The experience can be pleasant or unpleasant, but whichever it may be, the *jīva* has to "suffer" it due to its total attachment to the world at the time of the encounter. $\delta\tau$ is thus a measure of the "incremental suffering" of the *jīva*. On the other hand, if it were a fully detached *jīva*, i.e. $\theta = \pi/2$, then $PR = 0$ and $RQ = \delta G = \delta s$, i.e. the projection is fully on the vertical axis. Endowed fully with knowledge and detachment, this *jīva* merely witnesses its encounter with the world and therefore does not "suffer" the experience at P. Actions performed at point P also do not add to the *jīva*'s bondage, instead they purify the mind by exhausting the *vāsanā*s. As taught exhaustively in *Gītā*, this *jīva* grows in purity of mind and spiritual knowledge through *karma-yoga* (Swami Chinmayananda 1975). δG is hence a measure of the "incremental merit and knowledge" gained by the *jīva*. The vertical axis G itself can be viewed as the axis of "knowledge and goodness" where knowledge indicates the knowledge of self (*ātmajñāna*) and goodness indicates the universal, unconditional love towards all beings arising out of total detachment and knowledge of self.

For a *jīva* with spiritual detachment somewhere between zero and $\pi/2$, $\delta\tau$ and δG will be both non-zero. We can surmise that there will be some incremental suffering as well as some incremental merit accruing to this *jīva* due to its encounter at P with the world. The suffering will be less, and the incremental merit more, than that of a totally ignorant, fully attached *jīva*.

$\delta\tau$ can also be interpreted as the "incremental experienced time" (Chidambaram 2006). Time is experienced when a *jīva*, identifying with the world it encounters, experiences a changing world. Since $\delta\tau$ represents the extent to which the *jīva* is attached to the world during the interval δs, it serves to measure incremental time experienced by the *jīva*. In a later section in this chapter we will return to the topic of experienced time and how it differs from cosmic time and physical time.

As pointed out, the geometry in *fig*. 4.5 suggests the relation:

$$\delta S^2 = \delta G^2 + \delta\lambda^2 \qquad (1)$$

Let us compare this with the Minkowski's Metric, δd, for the four-dimensional time–space continuum defined by:

$$\delta d^2 = \delta S^2 + \delta t^2 \qquad (2)$$

where δS = the distance in the ordinary three-dimensional space, and δt = c. time, c being the speed of light. For convenience, the units of measurement of time and space may be set such that $c = 1$. Equation (2) reflects the mathematical relationship between time and space.

Physicists sometimes explain this basic relationship in a way that is particularly helpful to our discussions (Greene 2000). Everything in this universe can be thought of as constantly moving through the time–space continuum at the speed of light.[6] Even objects stationary in space are also travelling at that speed, except that their travel is in time dimension only. They are moving through time at the fastest speed possible, namely speed of light. We may say that time moves at the fastest rate for these objects. When something is

6 To talk about movement in space–time continuum may sound like a contradiction of the "static" universe. The timeless, static view is however strictly an "outside" view; observers within the universe do experience time and movement.

moving through space, it is in effect diverting part of its speed in the time dimension to accomplish travel in the three-dimensional space. Because of this diversion, its speed in time is reduced; that is "time slows down" for it. In the extreme case, time stops altogether if the object moves through space itself at the speed of light.

We re-write Eq. (2) as follows:

$$\delta d^2 = \delta S^2 + \delta i.t^2 \qquad (3)$$

where $i = \sqrt{-1}$, the unit imaginary number. Comparing equations (1) and (3), we see the correspondence between the following pairs of terms:

1. The four-dimensional time–space continuum *d* with the space $\mathcal{M}$;
2. Imaginary time *i.t* with experienced time τ on the horizontal axis; and
3. Three-dimensional Euclidean space *S* with *G*, the vertical axis representing goodness and knowledge of self.

Using this correspondence, the time–space relationship given above readily translates into a fundamental law of spirituality.

The Fundamental Law of Spirituality

Everything and every *jīva* is in the *Mahat* ($\mathcal{M}$-space) and all move constantly at a uniform speed across this space, just as everything in the material world is moving at the speed of light in the time–space continuum. In *fig*. 4.5, for example, the world, which is at arc *AB* at a certain time, is at arc *CD* after the elapse of δs units of cosmic time. All things and beings "move" in the $\mathcal{M}$-space with the universe. However, spiritual beings, or *jīvas*, by virtue of their detachment, divert part of this speed to advance in the dimension of knowledge of self and goodness along the vertical axis *G*. This reduces their speed in the dimension of "experienced time" or "suffering" τ (along the horizontal axis).

The speed component along the vertical axis, $\delta G/\delta s$, corresponds to the speed in space in the physical world. As seen in *fig.* 4.5, it equals sin θ which has been defined in the earlier referenced work (Chidambaram 2006) as the *jīva*'s "Relative Awareness", a comparative measure of the goodness and knowledge of a *jīva* relative to that of a fully realized "Perfect Being". The "goodness and knowledge" manifests in a perfect being as unconditional love for all beings, based on a full understanding of the oneness of all beings. Therefore, higher the vertical speed component of a *jīva*, higher is the degree of love and understanding it manifests, and lower is its horizontal speed component signifying worldly suffering. This leads to the "Fundamental Law of Spirituality" expressing the inverse relationship between love and suffering:

> As a *jīva* gains more and more in universal, unconditional love and understanding, it suffers less and less in the world.

This relationship, which in mathematical terms is identical to the relationship between time and space, can be justifiably called the *Fundamental Law of Spirituality* in view of the importance given in all spiritual and religious traditions to unconditional love.

Similarities between Vedānta and Relativity Physics

The many similarities between the material realm described by relativity physics and the spiritual realm as conceived in Vedānta can be now summarized as in Table 4.1.

The similarities are striking, but they gain even more significance in view of the fact that the terms "consciousness" and "space" are considered in Vedānta as synonymous in a philosophical sense. Similarly, "knowledge" is often equated with "light". This may be just a coincidence, or, may be the seers of the Upaniṣads did have an intuitive understanding of Einstein's time and space.

Table 4.1: Similarities between Material Realm and Spiritual Realm

Concept in the Material Realm of Relativity Physics	*Concept in the Spiritual Realm of Vedānta*
Space–time continuum, with coexisting past, present, and future	*Mahat* (field of experiences), with coexisting past, present, and future
Space (three-dimensional)	The vertical axis, *G*, of knowledge of self, and goodness (associated with self or consciousness)
Speed in space	Extent of establishment in self, or knowledge of self, given by sin θ, "speed" in *G*
Stationary in Space; zero speed in space*	sin $\theta = 0$, $\theta = 0$, totally attached to world, ignorant of self
Maximum speed in space, $c = 1$, is attained by light	Fully established state in self, sin $(\pi/2) = 1$, is attained by the enlightened (fully detached)
To attain speed of light, a particle must have no mass	To attain enlightenment, a *jīva* must have no attachment and no ego
An observer stationary in space is in the domain of time. But time comes to a stop for one moving at speed of light*	A *jīva* ignorant of self, is bound to time and suffers the changes that time brings. But an enlightened *jīva* goes beyond time and attains immortality
Space contracts to zero in the direction of travel for one travelling at speed of light. Thus, there is no further forward to go*	Everything is known by a *jīva* that realizes self. Thus, there is nothing more to be known

* Relative to some fixed frame of reference.

Cosmic Time, Physical Time, and Experienced Time

The distance between two consecutive waves in *fig*. 4.3 can be considered as a unit "cosmic time" since this is the interval between the cosmos "now" and the cosmos "next". The term "cosmic time" is somewhat arbitrary since, as we saw earlier, the interval between two waves is not time but more like the Minkowski measure of distance in space–time. There is no concept of time when viewed from a point outside the field of experiences of *fig*. 4.3, or outside the space–time continuum of *fig*. 4.4. There is no clock in the cosmos that can measure the cosmic time. But time *is* experienced by sentient beings within the space–time continuum or the field of experiences. The time measured by a sentient observer within the physical realm of space–time continuum depends on its motion in space and may be called "physical time" or "personal time". It is measured by a physical clock which is stationary relative to the observer. The time experienced by a *jīva* in the spiritual realm is yet another concept, different from both cosmic time and physical time. It depends on the *jīva*'s spiritual detachment. This "experienced time" is more mental or psychological and cannot be measured by a physical clock either.

We may illustrate the three different time concepts with the example of twin siblings sometimes employed in teaching relativity physics. John and Jay are twins who are together at time t_1 celebrating their 19th birthday. The same day Jay leaves for a distant galaxy, travelling at a very high speed. After reaching the galaxy Jay turns around instantly and returns to the earth at time t_2. Because of his high relative motion in space, physical time for Jay runs, let us say, 60 per cent slower than for John. Five years have passed on the earth when Jay returns just in time for the twins to celebrate their birthdays together again. But now there is a big difference: it is the 24th birthday for John and 21st for Jay. All this is shown in *fig*. 4.6 by plotting the locations of J_1 (for John) and J_2 (for Jay) at the two birthday party events. At time t_1, J_1 and J_2 are at the points P_1 and P_3 respectively, and at time t_2 they are at P_2 and P_4. Though J_1 and J_2 will be co-located in space-time continuum at the two birthday events, they are in different locations in the $\mathscr{M}$-space in view

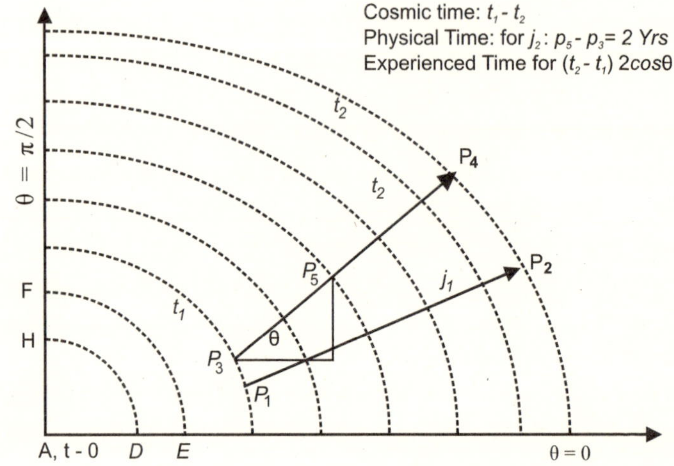

fig. 4.6: Cosmic, physical, and experienced time

of their different spiritual detachment. This underscores the fact that the $\mathscr{M}$-space is not the same as space-time continuum, though the two have similarities.

Cosmic time between the two birthday events is the interval (t_2-t_1) between the two waves. As emphasized earlier, we call this interval as time only for sake of convenience. The physical time corresponding to this cosmic time is different for the two brothers. For J_1, bound to earth, physical time elapsed is 5 years, whereas for J_2 it is 2 years. This difference is due to the effects of relative motion in space–time continuum. The cosmic interval (t_2-t_1) is itself neither 5 years nor 2 years. We cannot measure it in time. It is just that the interval P_2-P_1 in the J_1's trajectory translates into 5 years, but the interval P_4-P_3 translates into 2 years. We may show this disparity in *fig*. 4.6 by locating J_2 at a point P_5 which is 2/5 the distance from P_3 as P_4 is from P_3. It is as if J_2 has moved from P_3 to P_5 during the interval (t_2-t_1) when J_1 moves from P_1 to P_2. This representation however compromises the "outside" view of the cosmos, since a wave is not now associated with a definite cosmic time tag.

But it may be more suitable to describe a *jīva*'s "inside" personal view of the world it encounters.

The time we *experience* is a function of our mind and it is more psychological than physical. Clearly, it is not the same as the time registered by the watch we wear. In deep sleep, for example, we are not with the world nor are we aware of time, but the watch keeps ticking. More generally speaking, the degree to which a *jīva* is identified with the world (as given by the *jīva*'s spiritual detachment) is a determinant of its experienced time. In *fig*. 4.6, physical time is 2 years for J_2, but his experienced time, defined as the projection of the cosmic time on the horizontal axis, is $(t_2 - t_1) \cos \theta$. In the previously referenced mathematical model of spirituality, "experienced time" is a proxy for a *jīva*'s suffering in the world. The logical link between experienced time and suffering is provided by the fact that it is the changes with which a *jīva* identifies that give rise to both suffering and the notion of time. When one is able to rise totally above changes through spiritual detachment, there is neither suffering nor time.

There are now two questions of major significance which remain to be addressed:

1. How is that sentient observers (*jīvas*) experience time when the there is no such thing as time in creation *per se*?
2. If past, present, and future coexist, does it not deny the free will for the *jīvas* to change the future by what they do now?

Perception of Time

The concept of time remains an enigma to science to this day. But Vedānta, which views liberation as the passage from mortality (i.e. transient life) to immortality (eternal life), must necessarily have a good understanding of the meaning of time. The Vedāntic view of time is clearly expressed in the following statement of Śrī Ramaṇa (Sri Ramana Ashrama 2000):

> What is time? It posits a state, one's recognition of it, and also the changes that affect it. The interval between two states is called time.

It follows from this definition that, for time to be experienced, there must be changes and also there must be recognition of the changes by some cognizant being. Where there is no change, there is no sense of time. *Brahman* (Self), is changeless and hence it is timeless and immortal. On the other hand, from the perspective of *jīvas*, the world of objects–emotions–thoughts, is always in a state of flux. *Jīvas* experience the states of the world and perceive the interval between the experiences as time. The successive states of the world were shown in *fig*. 4.3 by a series of waves. A *jīva*'s encounters with the successive states of the world can be shown, as in *fig*. 4.7, as a point on each wave.[7] As the *jīva* moves from one

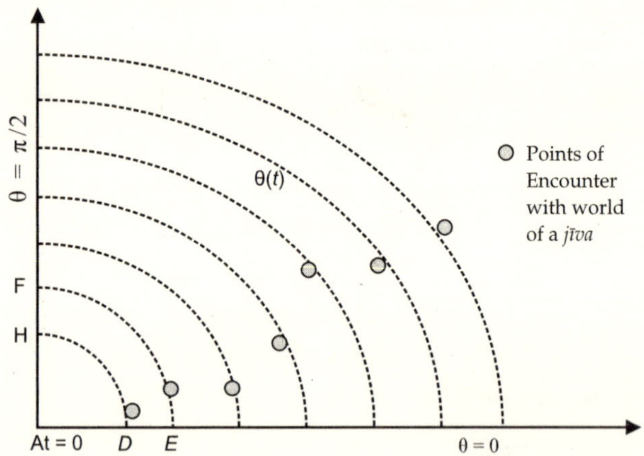

fig. 4.7: *Jīva*'s encounters with the world

[7] The location of the point on the wave depends on the *jīva*'s spiritual detachment. Since the wave as a whole represents the state of the world, the location does not affect *what* the *jīva* encounters. However, the spiritual detachment does determine *how* well spiritually the *jīva* handles the particular encounter.

wave to the next, it experiences the changing states of the world, and the experience of changes results in the perception of time.

What propels a *jīva*, one may ask, from one wave to the next? In other words, what keeps a *jīva* bound to time and to the changing world? From the point of view of science, worldly life is all that there is and hence the question is irrelevant. From a spiritual point of view, on the other hand, this is a very valid question. Vedānta says it is *māyā*, and the consequent ignorance (*avidyā*) that keeps the *jīva* bound to world by its power of delusion. It is the same *māyā* that *Īśvara* (the Creator), wields to project the world on the cosmic mind. The difference is that, *māyā* is under the control of *Īśvara*, whereas *jīva*s are controlled by *māyā*.

What Moves from Wave to Wave?

What exactly do we mean when we say a *jīva* moves from wave to wave? The answer Vedānta gives, as well as what some neuroscientists now suggest, challenges some of the deepest beliefs we have about ourselves and the world we live in.

According to Vedānta, a *jīva* has three bodies: a *gross* (physical) body, a *subtle* body constituted of mind and intellect, and a *causal* body constituted of its *vāsanā*s (inherent tendencies). All three bodies are part of the creation. Hence these bodies cannot be what moves from wave to wave as in *fig*. 4.7, to produce the life's experiences for the *jīva*. None of these three bodies is sentient and therefore none, by itself, can be the primary instrument of our experiences. Scriptures suggest that the sentient *jīva* is born when the *ātman* (self) "enters" the individual gross–subtle–causal body complex. That is, sentiency in an otherwise insentient body complex is produced by its association with the self, which is of the nature of consciousness. Some Vedāntins make this process a little bit more specific with the Pratibimbavāda (Theory of Reflected Consciousness) (Swami Swahananda 1967). They suggest that consciousness reflecting off the intellect (*buddhi*) produces the "beam of intelligence" in the light of which a *jīva* recognizes the world of objects–emotions–thoughts. According to Ādi Śaṅkara *buddhi* is the determinative faculty in us. An object is recognized when the *buddhi* determines what that object is.

The *jīva* cognizes and experiences the state of the world represented by a wave in the beam of reflected consciousness. This is what we mean by descriptions such as "*jīva* encountering a wave" or "*jīva* meeting the world". The wave and the state of the world it implies for the *jīva* have been there forever in the timeless $\mathscr{M}$-space as though patiently waiting to be illumined. A stream of experiences is produced as the beam of reflected consciousness illumines wave after wave. This stream of experiences is verily the *jīva*'s life and its sequential character is the source of the experience of time. In this respect, the waves are like tracks of music recorded on a CD and the reflected consciousness of the *jīva* is like the laser "read head" moving over the recorded tracks to make life come alive.

The Grand View and the Science to Support It

There are a countless number of *jīva*s in the cosmos, ranging from plants and animals to humans to angelic beings (*deva*s). Each *jīva* has a unique view of the world, its experience at any time being unlike that of any other *jīva*. A happy moment for one *jīva* can be a painful one for another. The grand picture that emerges is that of a multitude of *jīva*s moving from wave to wave in the *mahat*, each experiencing what is in store for itself. It is a picture that sharply differs from what we generally believe: That the future is yet to be born, that we create the future by what we do "now" and that what we do now is according to our will. The truth, we find, is that we experience only what is always existent and we can do nothing "now" that has not already been done through us. We are not the authors of what we do.

Science is also veering towards similar views in its effort to explain how there can be the dynamic experience of time and changes in a timeless, static universe. The explanation, as the following quotations from three scientists show, usually invokes consciousness.[8]

According to Russell Stannard (1987), (Physicist, Open University of UK):

> Physics itself recognizes no special moment called "now" — the moment that acts as the focus of "becoming" and divides the "past" from the "future". In four-dimensional space–time nothing changes, there is no flow of time, everything simply is. . . . It is only in consciousness that we come across the particular time known as "now". . . . It is only in the context of mental time that it makes sense to say that all of physical space–time is.

Says Hermann Weyl, 1922 (Mathematician-Physicist, Advanced Institute of Science, Princeton):

> The objective world simply is, it does not happen. Only to the gaze of my consciousness crawling upward along the lifeline [world line] of my body does a section of this world come to life as a fleeting image.

Roger Penrose, 1994 (Mathematician) observes:

> . . . particles do not even move, being represented by "static" curves drawn in space–time. Thus what we perceive as moving 3D objects are really successive cross-sections of immobile 4D objects past which our field of observation is sweeping.

---

[8] The quotations are taken from the previously referenced paper by J. Smythies (2003).

Consciousness figures prominently in the above statements. This is least surprising to a Vedāntin. After all, "consciousness is *Brahman*" (*prajñānam brahma*) is the first of the four "great statements" of the Vedas (Swami Chinmayananda 1982). Everything in creations arises, lives, and dissolves in that consciousness.

A Neuroscientist's View of Reality

Scientists, however, are not yet ready to accord the same supreme status to consciousness as Vedānta does. The field of cognitive science and neuroscience is a very active one today and teems with a number of different hypotheses about consciousness and its purpose. A full survey of this important body of literature is not within the scope of this chapter, but it is noted that a consensus is yet to emerge as to what consciousness is or whether it can be ever described satisfactorily in purely neurobiological terms. The situation, after decades of research, has been summed in one recent book thus:

> Only one proposition about how the brain makes us conscious . . . has emerged unchallenged: we don't have a clue.
>
> — Noe 2009

This pessimistic assessment is not meant to overlook the steady progress being made by neuroscientists on many fronts, but it does underline the significant challenges encountered in studying consciousness objectively, when in fact, consciousness is very much subjective.

The current prognosis can be discouraging to those who hope to explain consciousness in terms of physical brain activities. But other scientists who approach the study of consciousness with an open mind are not daunted by the negative results. For example, the views of John Smythies, a neuroscientist from University of California at San Diego, are illumining and very relevant to our discussions. In his previously referenced paper (Smythies 2003) on "Time, Space, and Consciousness", Smythies makes three key points:

1. What we perceive is only a representation of the external world constructed by the brain and not the external world itself. There is a "physical world" out there and a corresponding, but separate, "phenomenal" world inside. To quote the author:

   > Phenomenal objects, together with the space in which they are located, are constructs of the central nervous system and in no sense are they direct views of the external objects that they represent. . . . The results of a large number of experiments in psychophysics . . . demonstrate beyond any doubt that, in vision, we do not perceive the world as it actually is, but as the brain computes it most probably to be.

2. The phenomenal space, or the space in which the phenomenal objects are located, is different from the four-dimensional space–time of physics. Smythies refers to the eleven-dimensional space and "branes" proposed by string theorists and suggests that the "phenomenal space and physical space are simply different spaces, different parallel universes whose contents are causally related".[9] That is, the phenomenal space lies on a multi-dimensional "brane" separate from the physical four-dimensional space–time. Smythies does not define or explain consciousness, but suggests that "consciousness is in this 'brane' and not in the brain". That is, consciousness is associated with the mental or phenomenal world and not with the brain which is part of the physical world.

---

[9] "Brane" is the short form for membrane and is a multi-dimensional space or sub-space. Goswami, in the previously referenced work (Goswami 2001), also considers the physical and mental worlds to be correlated, parallel quantum systems.

3. The physical and phenomenal worlds, though in separate dimensions, are causally related and therefore affect one another. The physical space–time is static with coexistent past, present, and future. But the phenomenal world is dynamic since there is perception of changes. Smythies suggests that the perception of time is generated by the motion of the phenomenal space relative to the physical space. The "now" is the present point of contact between the two spaces. Compare this with the Vedāntic model where the "now" is the point of contact between *jīva*, the experiencer, and world, the experienced.

The general picture that Smythies paints of reality is similar to the view presented in *fig*. 4.3. This is remarkable given that the starting point for *fig*. 4.3 was the Vedāntic teachings, whereas Smythies bases his views on relativity physics and modern neuroscience. It is noteworthy that a neuroscientist with intensive clinical experience should admit that mind is not the brain and propose that consciousness and mind are in a space altogether separate from the physical space of the brain. Also, his proposal that the "now" is the point of contact between the physical and phenomenal spaces is similar to the Vedāntic model (*fig*. 4.7) where "now" is the point of contact between the sentient *jīva* and the insentient O-E-T. The physical and phenomenal worlds correspond to the *virāṭ* and *hiraṇyagarbha* respectively of Vedānta (Swami Chinmayananda 1953).

There are differences, however. Smythies suggests that consciousness is in the phenomenal space or "brane". Vedānta denies absolute reality to both the physical and phenomenal world, consciousness being the only reality. Therefore consciousness cannot be in the brane of phenomenal space; rather the brane, as well as the physical space is in *mahat*, the cosmic mind, which itself is in consciousness. Secondly, there is a question about the nature of the "brane" and the individual observer's phenomenal space. Vedānta asserts that there is only one consciousness but with many individual reflections. Smythies seems to suggest that there is one brane in which there are many individual "consciousness modules". A final significant difference is in regards to the spiritual dimension of our existence. Smythies, keeping his discussion within the traditional confines of science, does not talk about spirituality. However, if reality is the timeless, changeless universe physicists say it is, and if individual experience of time is (to use Einstein's words) only an illusion, then should not awakening from this illusion to that timeless reality be a goal of all individuals, especially when the cause of suffering is this illusion? This in fact is the goal of spirituality as repeatedly stated in Upaniṣads. However, neither in the four-dimensional space–time continuum of the physicist, nor in the parallel universe extension proposed by Smythies, is there any possibility for a *jīva* to go "outside" the confines of their respective domains to attain timelessness. Only in the Vedāntic model there is the possibility for *jīva* to reach immortality by increasing detachment.

We now turn our attention to this goal and the role of free will in achieving the goal.

Free Will

In the cosmic scenarios shown in *figs*. 4.3 and 4.4, the world of the future coexists with the present. Hence, what a *jīva* will meet at the "next" moment is beyond its power to change by what it may do "now". This basically denies any role for personal free will in our life. If a *jīva* has personal free will, it can exercise that will to engage now in actions that would change the world at the next moment. Since this possibility is negated, one must also negate, or at least greatly restrict, the free will attributable to *jīva*s. No doubt, denial of personal free will is very unpalatable to our usual way of thinking, but, for reasons discussed below, a fact nevertheless.

It is not uncommon for us to sometimes surprise ourselves by what we do. Our act may be contrary to what we generally believe in and we may never quite understand why we did what we did. There are also situations where we have to make a decision quickly without having any time to reason it through. This shows that there are instances when our decisions are not necessarily dependent on our conscious will. But what may be surprising is to be told that our decisions even in other cases are not ours. Yet our scriptures, logic, science, and seers all seem to proclaim this as the truth.

Scriptures: Akartāham (I am not the doer) is a phrase heard again and again in Vedānta. In the *Bhagavad-Gītā,* the Lord calls those who consider themselves as the doers of their actions as "deluded", reminding us that all actions are in fact accomplished by the forces of *prakr̥ti.*

prakr̥teḥ kriyamāṇāni guṇaiḥ karmāṇi sarvaśaḥ |
ahaṁkāra-vimūḍhātmā kartāham iti manyate ||

— *Gītā* III. 27

Elsewhere in the *Gītā,* we are also told that *Īśvara,* seated in the "heart" of all things and beings, controls them.

īśvaraḥ sarva-bhūtānāṁ hr̥ddeśe 'rjuna tiṣṭhati |
bhrāmayan sarva-bhūtāni yantrārūḍhāni māyayā ||

— *Gītā* XVIII. 61

Thus, the notion that what we do is a result of our individual will, ignores the truth that all actions issue out of the forces of *prakr̥ti* and Will of *Īśvara.*[10]

---

10 This does not deny that there is free will affecting the evolution of the world. There is free will but it belongs to the *Īśvara*. Vedānta views *Īśvara* as the supreme controller of all things and beings. *Īśvara* exercises the divine will while remaining within the bounds of *prakr̥ti*'s laws. This is the *Prakr̥ti–Īśvara* paradigm. For a full discussion, see Chidambaram 2004.

Logic: There is a logical fallacy behind the presumption of personal will. This is highlighted by Einstein when he pointedly asks *"Do we will to will?"* The implication of the question is that willing is not a conscious action. Any conscious action requires a conscious decision to perform the action, otherwise it is a spontaneous, unconscious or reflexive action. Therefore, if willing were a conscious action, we should have consciously willed to will. If we willed to will, the same logic suggests that we should have "willed to will to will" and so forth, *ad infinitum*. But this process cannot go endlessly and should end somewhere in a spontaneous "unwilled" decision. Willing, we therefore conclude, involves an "unconscious" decision arising somewhere deep inside us. In other words, *willing is not really something we do, but something done to us*. This place "deep inside us" is what is indicated as the "heart" in the Upaniṣads and in the above quotation from the *Gītā*.

Neuroscience: Experiments done in 1977 by Dr. Libet of University of California, demonstrated that choices made by subjects were registered by the brain half-a-second *before* the subjects were aware that they have made a choice. Similar experiments repeated more recently using state-of-the-art fMRI techniques have produced evidence of even longer delays before subjects become aware of their choices. The results are consistent with the view that we are not the real authors of our decisions.[11]

Spiritual Evolution of Jīvas

Does it mean that *jīva*s have absolutely no free will? That is not the case, according to Maharṣi Śrī Ramaṇa, the great spiritual master of our times. Śrī Ramaṇa emphasizes that

[11] Our common belief is that what we do follows from what we decide to do. That is, what we do is the effect and what we decide to do is its cause. The results of the Libet experiments suggest that this may not be the case. Both the action and the intent to act are effects of brain states which are determined as per the *Prakṛti–Īśvara* paradigm.

everything that happens in the world happens only at the will of *Īśvara* (Mudaliar 2006). A *jīva* must submit to that will and accept whatever is ordained by *Īśvara*. But, adds Śrī Ramaṇa, the *jīva* has one precious freedom to exercise its personal will. The *jīva* has no control over *what* it meets in life, but it has control over *how* it meets life. Basically it has two options at any given time: (1) Identify with the world and "eat" (i.e. suffer or enjoy) the experience, or (2) Identify with the self and be a mere "witness" to the experience. This choice presents itself every moment of our life and our choice will determine whether we advance in our knowledge of self or merely prolong our suffering in the world. In effect Śrī Ramaṇa teaches us that, while our material life is totally controlled by *Īśvara*, our spiritual destiny is something we can and must control. Indeed it is our bounden duty as a sentient being (*puruṣārtha*) to do so.

The spiritual detachment of the *jīva* at time t, denoted by $\theta(t)$, determines which of these two options the *jīva* is likely to choose at time t. In the proposed mathematical model of spirituality, it is shown that the probability that a *jīva* with spiritual detachment $\theta(t)$ will choose to identify with the Self is $\sin^2\theta(t)$ and the probability that it will identify with the world is $\cos^2\theta(t)$ (Chidambaram 2008). As noted earlier, $\theta(t)$ itself depends on the *jīva*'s *vāsanās*. So, in effect, the *vāsanās* determine the likelihood of choosing either option. The actual option chosen depends on the *jīva*'s will-power. For example, at a moment where tāmasic *vāsanās* dominate, spiritual detachment θ and $\sin^2\theta(t)$ will be small, and hence it is much more likely that the *jīva* will identify with the world and not the self. However, we cannot totally rule out the latter possibility. Given sufficient will-power, even this *jīva* can overcome the pull of the lowly *vāsanās* and identify instead with the witnessing self. Will-power is a primary requirement for spiritual progress and its exercise is the (only) prerogative of the *jīva*.

The function $\theta(t)$ describes the spiritual evolution of a *jīva* over time. This function takes values in the range $(0, \pi/2)$, with the convention that at the beginning $t = 0$ every *jīva* is in delusion;[12] i.e. $\theta(0) = 0$. The life's experiences gradually bestow knowledge on the *jīva*, giving it increased detachment until, in some future birth, the *jīva* attains total detachment (i.e. $\theta = \pi/2$) or Realization. Plotted in the $\mathscr{M}$-space, the function is a curve starting from the origin and ending on the vertical axis at the time of realization (see *fig.* 4.8).

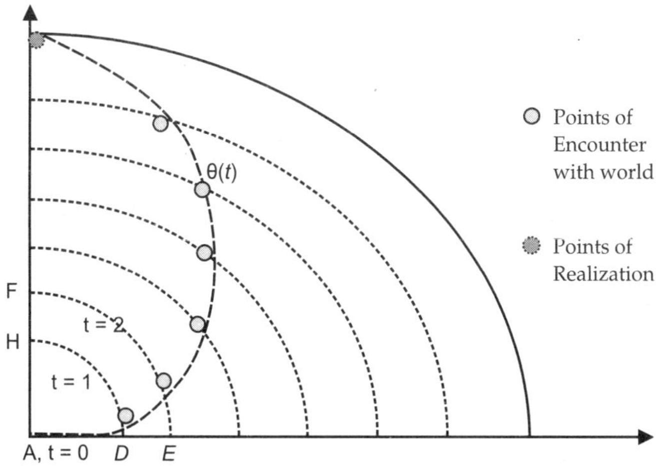

fig. 4.8: Path of spiritual evolution in the $\mathscr{M}$-Space (polar co-ordinates)

The CD Analogy

As the above discussions show, both Vedānta and modern science suggest a picture of the reality that is unlike anything we generally take to be the truth. A good, though not perfect, analogy to describe that picture happens to be an audio CD and the way it is played back. The hundreds of tracks burned on the disc encode music. When the CD is played (that is, when

[12] Actually, at $t = 0$, the "wave" is a mere point and the value of $\theta(0)$ is therefore irrelevant. This makes spiritual sense: At $t = 0$, the world is yet to emerge and hence the question of detachment from world has no meaning.

the laser head reads the information coded on the tracks and converts it to audible sound), the music comes to life. We hear the beginning notes followed by the subsequent notes until the final notes are sounded. We hear the beat of the music and sense the movement in the melody. Altogether, when the CD is played back, there is vivid experience of time. But if we were to just pick up and look at the CD we will not hear the melody nor feel the beat of music which lies hidden in it.

Similarly, the waves in the $\mathscr{M}$-space are records of the state of the world from the beginning of the creation till its end, and include the past, present, and future experiences of every *jīva*. From the point of view of the creator there is no sense of time associated with $\mathscr{M}$-space or creation. The sense of time is experienced by a *jīva* when the light of its (reflected) consciousness illumines wave after wave to bring to life the particular experience recorded in it. Thus does a *jīva* proceed from experience to experience in its life.

The "fly-on-the wall" description of the *jīva* is also an apt one. The wall (namely, the *mahat*) displays all the waves of creation at the same time. But a flat, two-dimensional fly crawling along the wall can see only the wave it is on. It does not see the wave "behind" nor the wave ahead. That is, it does not see the past or future, but only the present. Only a Creator who is not on the plane of the wall has the view of the entire wall and creation. This analogy is reminiscent of Puruṣa-Sūktam where the Creator is said to be "ten digits" above the creation (Swami Chinmayananda 1999).

That the entire field of experiences is confined to the first quadrant of the two-dimensional (t,θ) space also reminds us of another *mantra* in the same Puruṣa-Sūktam:

> *etāvānasya mahimāto jyāyām̐śca pūruṣaḥ* |
> *pādo 'sya viśvā bhūtāni tripādasyāmr̥taṁ divi* ||
>
> — *R̥gveda* X.90.3

> All creation is His (Creator's) Glory, but He is greater than these still. The entire creation is but His one quarter, the other three quarters are immortal, divine.

In our mathematical representation, the second, third, and fourth quadrants are beyond the sensory experience of a *jīva*. Instead of knowing itself to be the illuminating principle that allows the recorded experience come alive, when a *jīva* takes the worldly experience to be its own, it "suffers" the consequent pain or pleasure. However, the *jīva* does slowly gain knowledge and increased detachment by the lessons learnt from worldly suffering, until, at long last, it matures in knowledge and attains realization.

A Testable Hypothesis

In this chapter the authors have proposed a new model that integrates Vedānta with modern science. But can one test and verify any of the model predictions? The Fundamental Law of Spirituality stated earlier in this chapter offers a hypothesis that can be potentially subjected to scientific testing and verification. The law states that a *jīva* suffers less and less in the world as it gains more and more in goodness and knowledge. In Vedānta, the last two terms are often given the specific meaning of "being able to see oneself in others, and others in oneself". Such a broad compassionate vision is the basis for the universal, unconditional love that a saint has towards all beings in the creation. Closer a person is to this ideal, the more contented is his or her life. All religions teach essentially the same truth but may differ in language employed. The biographies of many great saints also illustrate this truth by recounting how joyfully such men and women lived, often in the face of tremendous physical and social adversities.

This spiritual truth can be brought within the ambit of modern science by experiment and verification. Note that the Fundamental Law of Spirituality does not directly involve concepts of god, heaven or consciousness any of which can quickly render a scientific study out of question. We are instead focusing on attributes of individuals such as detachment, suffering, compassion and selflessness. Though these are also subjective terms, they are within the measurement and analytical capabilities of present-day psychometric and neuroscience disciplines. For example, psychological and neuro-imaging studies have been used to study empathy, an attribute related to universal, unconditional love (Lamm et al. 2007).

Psychometric instruments are used today in measuring such characteristics as "religiosity" of an individual and their effect on daily life (Underwood and Teresi 2002). The National Institute of Health has funded, and continues to fund, studies of this nature in exploring the effectiveness of "alternative medicine". We believe that psychometric instruments can be constructed, or even adapted from those already used in other studies, to assess an individual's

(a) level of contentment in daily life (life satisfaction scale).

(b) degree of selflessness and detachment. This is the opposite of narcissistic tendencies and dependence.

When administered to a large enough sample of individuals in a statistically designed study, the results can be used to see how the level of contentment an individual feels correlates with his or her attributes of love, selflessness, etc.

Techniques currently used in neurosciences are also relevant in studying, and possibly measuring, spiritual detachment. Spiritual detachment, according to our theoretical construct, specifies the *jīva*'s probability of staying in the witness mode in the face of sensory and psychological stimuli flowing into its consciousness from the outside world and from its own mind. fMRI techniques have advanced to a stage where it seems possible to determine when a subject is in the witness mode, rather than in the indulging mode, by looking at the brain regions activated. In the fMRI and PET scanning literature, two modes are being recognized: (1) Attentional engagement mode and (2) Resting default mode (Raichle and Snyder 2009). In this manner one can evaluate an individual's detachment and then correlate it with results of psychometric tests assessing the individual's degree of contentment in daily life. Theory suggests that we can expect individuals with higher levels of detachment to lead a less troubled, happier life.

The Importance of Research and Education in Spirituality

Spiritual health is at least as important as physical health in determining an individual's quality of life, including his or her ability to live harmoniously with others in the society. A proper spiritual perspective can go a long way in bringing about not only happiness to the individual but also peace in the community. The world today is beset by conflicts of all sorts — racial, economic, nationalistic, and religious — which threaten to tear us apart. Spiritual education is one of the most cost-effective strategies we have to combat these negativities with positive values.

Yet, education and research in spirituality languishes due to want of support and is thus prevented from delivering the huge benefits it is capable of. A primary reason for this indifference is the mistaken identification of spirituality with religion. It is true that historically religions have been the primary, and often only, source for spiritual education. Yet, it is also true that spirituality transcends religion and can be pursued regardless of one's religious beliefs.

Further, it is becoming increasingly clear that spirituality can be approached and studied as a science in its own right (Chidambaram 2009). This holds the prospect that there may be a day in the not too distant future when spiritual education is offered in schools and universities as part of their science curriculum, much as biology, chemistry and medicine is today. But it will surely take a good deal of foresight, wisdom, persuasion and co-ordinated effort at all levels of the society. The future of the human race could depend on whether or not we mount and succeed in such an effort.

Appendix

Response to Reviewer's Comments/Questions which were/are not addressed in the paper since they are not considered important for the main topic discussed, namely Time

Q.V.18: If such a *jīva* is in an indulging mode, is he not experiencing the world? . . . How can then θ be 0?

A: Whether we experience the world or not, or what we experience in the world is not affected by θ. It affects only the extent to which we "suffer" from the experience. When $\theta = 0$, we are totally identified with what we experience and hence "suffer" the joys and sorrows of the experience most.

Q.V.19: Does not a perfected *jīva* witnessing the world suffer from a dualistic mentality of subject–object?

A: The subject–object mentality ceases only on *samādhi*, total absorption in self. There is no transaction with the world while in this state. A perfected *jīvanmukta*, like Śrī Ramaṇa, can transact with the world in witness mode (*sākṣī bhāva*). The subject–object duality is there but will not cause any suffering or bondage to this *jīvanmukta*.

Q.V.23: How do you know that Creator has the view of Creation shown in *figs*. 4.2 and 4.3. Can anyone know and have a conscious view of the whole "wall"?

A: Here, as well as in many other places in the paper, we are presenting the Vedāntic model and not the personal observations of the authors. Vedānta clearly states that *Īśvara* is beyond time, time and space being part of what is created.

Can anyone have the same view as the Creator? No, but it is also said that there are states of consciousness attaining which a *yogī* becomes the knower of past, present, and future (*trikālajñānī*.)

Q.V.24: Are creator and *mahat* the same?

A: No. As we have understood it, *mahat* is more like the Creator's mind.

Q.V.26: Define cosmic time.

A: Mathematically, it is the parameter *t*. As made clear subsequently that it is not strictly time, but more like the space-time metric.

Q.V.29-30: Is such a *jīva* attached to the bliss of *samādhi*? Does he have a sense of individuality?

A: Again, as in V.19 above, the reference here is to a *jīvanmukta*, an enlightened soul which nevertheless is still participating in the world. Such a *jīva* is not always in *samādhi*, yet enjoys the bliss that comes from total detachment (*kasya sukham na karoti virāgaḥ*? asks Ādi Śaṅkara in *Bhaja Govindam*). Identifying with all things and beings in the universe, its sense of individuality is nearly all gone.

Q.V.31: Define imaginary number.

A: The definition is given mathematically in the text, namely $i = \sqrt{-1}$.

Q.V.32: Explain goodness and knowledge of self.

A: This explanation has been added to text in the earlier section titled "Minkowski Metric. . . ."

Q.V.49: How about the *Gītā* VI.5.

A: The *Gītā* indicates in this *śloka* that the effort to spiritually uplift an individual must come from oneself. As explained in section 16 of this chapter, this is the only freedom of will the individual *jīva* has.

Q.V.50: Other recent works in neuroscience re: free will.

A: This is a valuable suggestion. I have not had time yet to get my hands on these references. I would appreciate if the reviewer can send us the link to online versions, if available.

Q.V.53: What happens when rājasic and sāttvic *vāsanā*s dominate?

A: The parameter θ signifying detachment will have higher values that when *tamas* dominates.

Q.V.54: Is this (future birth) scientific? Please avoid terms in a scientific article.

A: This is where we beg to differ. The paper presents the Vedāntic views and concepts. These should not be subjected to censure simply because science has not yet come around to grips with the issues. If science has definitely disproved a standpoint, then we do agree we should avoid it.

Q.V.60: Please define and differentiate spirituality, science, and religion.

A: It may be too much of a digression to incorporate these ideas into the present paper. The definition and differentiation is the subject matter of the referenced paper by Raju Chidambaram on "Science, Spiriutality, and Religion" which was also published in the September 2009 issue of Bhavan's journal.

References

Callender, Craig, 2010, "Is Time An Illusion?", *Scientific American*, June, pp. 58-65.

Capra, Fritjof, 1975, *The Tao of Physics*, Berkeley, CA: Shambala Publications.

Chidambaram, Raju, 2009, "Science, Spirituality and Religion", Third SSEASR Conference, Denpasar, Bali, Indonesia, June*.

———, 2008, "Sri Ramana's Principle of Personal Free Will", Seventh International WAVES Conference, Orlando, FL, June.*

———, 2006, "Towards A Mathematical Theory of Spirituality Based on Advaita Vedānta — Parts I and II", Sixth International WAVES Conference, July, Houston, TX*.

* These papers have been incorporated into the recent book *Law of Love and the Mathematics of Spirituality*, Raju Chidambaram (2011).

———, 2004, "Unraveling Uncertainty", Fifth International WAVES Conference, July, Washington, DC.*

———, 2008, "Vasana Functions and the Dynamics of Spiritual Evolution", Seventh International WAVES Conference, Orlando, Fl, June.

Chinmayananda, Swami, 1982, *Aitareya Upaniṣad*, Bombay: Central Chinmaya Mission Trust, Ch. 3 , v. 3.

———, 1953, *Discourses on Māṇḍūkya Upaniṣad with Gauḍapāda's Kārikā*, Bombay: Central Chinmaya Mission Trust, p. 21.

———, 1992, *Self Unfoldment*, Piercy, CA: Chinmaya Publications, p. 37.

———, 1975, *The Holy Geeta*, Bombay: Chinmaya Mission Trust.

———, 1999, *The Universal Person: Purusha Sooktam*, Central Chinmaya Mission Trust, Mumbai.

———, 1988, *Discourses on Mundakopanishad*, Ch. 3, Sec. 1, verses 1 & 2 Bombay: Central Chinmaya Mission Trust, rpt.

———, 1978, *The Crest Jewel of Discrimination: Comments on Adi Shankara's Vivekachudamani*, Bombay: Central Chinmaya Mission Trust, v. 497.

Goswami, Amit, 2001, *The Physics of the Soul*, Charlottesville, VA: Hampton Roads Publishing.

———, 1993, *The Self-Aware Universe: How Consciousness Creates the Material World*, New York: Tarcher/Putnam.

Greene, Brian, 2000, *The Elegant Universe*, New York: Vintage Books, pp. 49-50.

———, 2004, *The Fabric of the Cosmos*, NY: Random House.

Jitatmananda, Swami, 1986, *Modern Physics and Vedānta*, Mumbai: Bharatiya Vidya Bhavan.

Lamm, C., C.D. Batson & J. Decety, 2007, "The Neural Substrate of Human Empathy: Effects of Perspective-Taking and Cognitive Appraisal", *Journal of Cognitive Neuroscience*, 19: 42-58.

Mudaliar, A. Devaraja, 2006, "God and Destiny", *The Maharshi*, **16**(1): 4-5, Arunachala Ashrama, Jamaica Estates, NY, January/February.

Noe, Alva, 2009, *Out of Our Heads: Why You Are Not Your Brain and Other Lessons From the Biology of Consciousness*, Hill and Wang (As quoted in a review of the book in *Scientific American*, March 2009.)

Raichle, M.E. and A.Z. Snyder, 2009, "Intrinsic Brain Activity and Consciousness", in S. Laureys and G. Tononi (eds.) *Neurology of Consciousness: Cognitive Neuroscience and Neuropathology*, Burlington, MA: Elsevier.

Smythies, J., 2003, "Time, Space, Consciousness", *Journal of Consciousness Studies*, **10**(3).

Sri Ramana Ashrama, 2000, *Talks with Sri Ramana Maharshi,* California: Inner Directions Publishing, p. 454.

Swahananda, Swami, 1967, *Pancadasi of Vidyaranya Swami,* Madras: Sri Ramakrishna Math, Ch VIII.

Tejomayananda, Swami, 1998, *Yoga Vasistha Sara Sangrahah,* Mumbai: Chinmaya Mission Trust, p. 81.

Todeschi, Kevin, 1998, *Edgar Cayce on the Akashic Records,* Virgnia Beach, VA: A.R.E. Press.

Underwood, Lynne and Jean Teresi, 2002, "The Daily Spiritual Experience Scale: Theoretical Description, Reliability and Exploratory Factor Analysis", *Annals of Behavioral Medicine,* **24**(1).

5

Consciousness Normalization

Ravi Jata

Abstract

Science and technology have been playing a major role in the advancement of human civilizations. The civilizations developed from inscribing on stones to documenting in computers, talking to the next door to watching people in the space and from riding on bull-carts to flying in spaceships. Science has become an intricate part of the human day-to-day lifestyle. The human brain has made the species pretty unique when compared to other animals on the Earth.

Even though lifestyle advanced with science, individuals still struggled with their way of living. Irrespective of technology, medicine, various other comforts in life, people are still unhappy and continue to lead a stressful and sometimes unhealthy life.

It appears as civilizations are moving towards technological sophistication the unhealthy habits are also increasing proportionately. The conveniences provided by the modern societies, because they are just technological by nature, are not completely useful in helping humans in leading a healthy life. In fact, because of these modern attractions, humans are becoming lazier, unhealthier and often-times leading an unhappy life.

It appears there are two sides to the human civilizations. One side is science and technology which brings innovation, convenience into human societies, and the other side is the spirituality (as a way of living) which addresses the desire for peace and tranquillity inside individuals.

As science tries to find the reason and logic behind everything, spirituality will bring in the supernatural concepts and argue that they are things beyond science. More often than not these two sides ridicule each other and create a general impression that science and spirituality are two contradicting sides that don't blend well with each other.

This work proposes a new way of living that is completely logical, scientifically compatible and can be practised at all the time. The methodologies proposed are not only logical, completely explainable but they also provide a simple and the minimum set of principles that explain how to lead a normal, relaxed and peaceful life.

Hypothesis

The fourth state or the way of living proposed is the most normal, logical way of living that the consciousness should be in while performing regular activities in life. A small deviation from this state will put unwanted stress on its host, the human body.

Testability

There are various modern scanning techniques available to monitor the physiological changes in the body and the brain. Using the hypothesis, the most stressless body and completely rested brain-activity can be achieved. The methods proposed here, can train the humans to reach the fourth state, which is the most relaxed state, and the state of the human body can be verified using the modern scanning techniques like EEG, PET scan or fMRI.

Sanātana Dharma is Science

Sanātana Dharma is considered to be a way of living than just a religion. Sanātana Dharma literally means "eternal law" because as this work explains, the *dharma* is nothing but science. As physics fundamentals are applicable on any planet or in any solar system, the *dharma* is also applicable anywhere.

Scientific Approach

In this report, Sanātana Dharma will be interpreted using advanced theoretical physics. The new interpretations will demonstrate that the *dharma* scriptures are not just philosophy as traditionally believed for thousands of years, not just some religious or baseless works, but as a great way of living life, designed around advanced concepts of physics.

Paramātmā as Four-dimensional Continuum

This work will try to convince the reader that Sanātana Dharma defined the visible three-dimensional world as the four-dimensional continuum. The following sections will clearly show that the *dharma* tried explaining these advanced physics concepts to a common man using various methods.

In order to bring this advanced science to a layman's level the *dharma* designed an entire way of living around this concept. Instead of identifying a fictitious, imaginary godhead, the *dharma* explained this four-dimensional continuum in which we are living as *paramātmā* and tried explaining the true nature of the universe, using various simple methods.

The interpretation starts with Puruṣa-Sūkta (Chanana 1983), the hymn 90 mentioned in the tenth *maṇḍala* of the *R̥gveda*. Puruṣa-Sūkta *śloka*s appear to explain the *puruṣa* or *paramātmā* as the four-dimensional continuum.

fig. 5.1: Continuum and Puruṣa-Sūkta

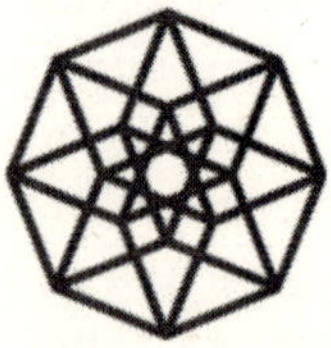

fig. 5.2: Tesseract in the Upaniṣads

सहस्रशीर्षा पुरुषः सहस्राक्षः सहस्रपात्।
स भूमिं विश्वतो वृत्वात्यतिष्ठद्दशाङ्गुलम्॥1॥

sahsraśīrṣā puruṣaḥ sahasrākṣaḥ sahasrapāt ।
sa bhūmiṁ viśvato vr̥tvātyatiṣṭhaddaśāṅgulam ॥1॥

— *R̥V* X.90.1

तमेकनेमिं त्रिवृतं षोडशान्तं शतार्धारं विंशतिर्पत्यराभिः।
अष्टकैः षड्भिर्विश्वरूपैकपाशं त्रिमार्गभेदं द्विनिमित्तैकमोहम॥4॥

tamekanemiṁ trivr̥taṁ ṣoḍaśāntaṁ
śatārdhāraṁ viṁśatirpatyarābhiḥ
aṣṭakaiḥ ṣaḍbhirviśvarūpaikapāśaṁ
trimārgabhedaṁ dvinimittaika-mohama ॥4॥

— *ŚvetUp.*

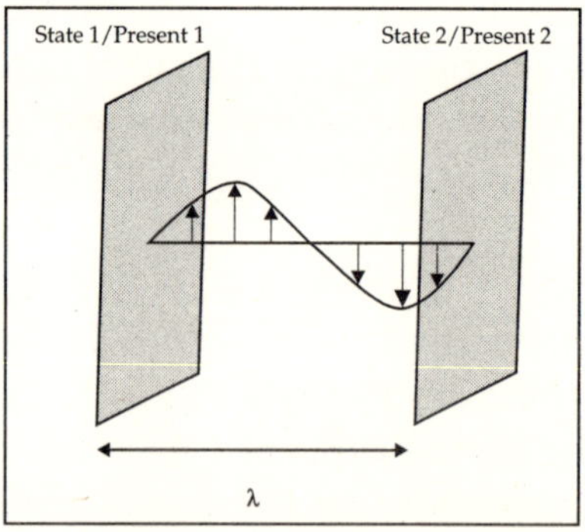

fig. 5.3: Cyclic Universe

In modern physics the four-dimensional mathematical space is represented by tesseract (Coxeter 1973; Gardner 2001). This chapter will also connect the features of the tesseract with the explanation of *śloka* 4 in *Śvetāśvatara Upaniṣad* (Tyagisananda 1986).

Fourth Dimension: New Concept

Sanātana Dharma can be understood by the advanced theoretical physics concept which is not yet realized by the current science and technology. In this chapter, for the first time, a new fourth dimension concept is proposed to completely define the three-dimensional world in which we are living.

The fourth dimension is derived from the fractional existence of the universe at any given time. Considering the past and the future are only creations of human mind and it is the present that is the reality; the fourth dimension is defined based on the reality that how little the state of present can span, thereby, if the present is only real then how fractional is the existence of the universe, is discussed. Using this new concept, the Sanātana Dharma scriptures are interpreted.

Tesseract

HIGHER-DIMENSIONAL REPRESENTATION

IN this section, we will discuss about how the four-dimensional space is represented mathematically. Physically, it may not be possible to sense the higher dimensions and sometimes they are only hypothetical or imaginary.

Just like the three-dimensional space is represented by a cube, the higher-dimensional spaces, higher than the three-dimensional space, are represented by more than one cube. For example, if the dimension of a mathematical space is N, then the number of cubes used to represent the space is:

$$C_n = (N\text{-}3) + 1 \text{ for all } N > 3,$$

where C_n = no. of cubes and
N = dimension of mathematical space.

So if $N = 4$, i.e. a four-dimensional mathematical space, then the dimensional space is represented by (4-3) + 1 = two cubes.

Even though the representations are multi-cubed, the cubes are "combined" to make them as one entity. The resultant of this single entity is called a hypercube, *n*-cube (Coxeter 1973).

FOUR-DIMENSIONAL REPRESENTATION

The four-dimensional mathematical space is represented by combined resultant of two cubes, called as tesseract (Coxeter 1973; Gardner 2001). In geometry, the tesseract, also called 8-cell or octachoron, is the four-dimensional analogue of the (three-dimensional) cube, where motion along the fourth dimension is often a representation for bounded transformations of the cube.

A tesseract is in principle obtained by combining two cubes. The scheme is similar to the construction of a cube from two squares.

Dimensional representation is created by connecting two prior dimensional representations, i.e. by connecting two dots a line is created. By connecting two lines a square is created and by connecting two squares a cube is created.

Similarly, put side by side two copies of the lower dimensional cube and connect the corresponding vertices. Each edge of a tesseract is of the same length. A multitude of cubes that is nicely interconnected. Two parallel cubes *ABCDEFGH* and *IJKLMNOP* can be connected to become a hypercube, with the corners marked as *ABCDEFGHIJKLMNOP*. (see *fig*. 5.4).

TESSERACT PROPERTIES

Summarizing the tesseract features:

SIXTEEN CORNERS

There are sixteen corners or vertices in the tesseract. Eight vertices per each seed cube. In *fig*. 5.4, these corners are marked as *A-B-C-D-E-F-G-H* and *I-J-K-L-M-N-O-P*.

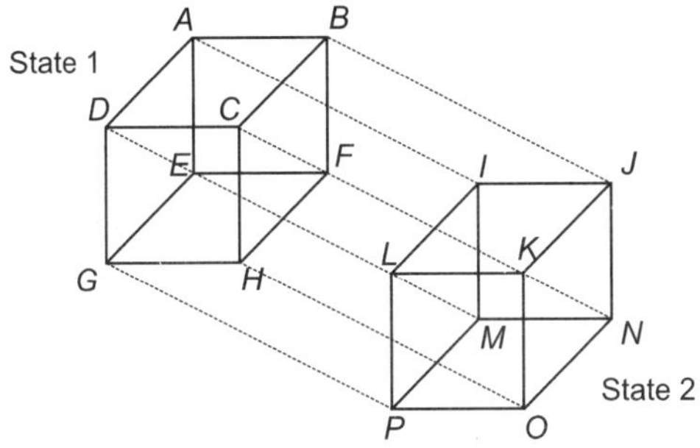

fig. 5.4: Tesseract

FORTY-EIGHT FACES

There are eight resultant cubes in the tesseract. Considering six faces per cube, there are a total of forty-eight faces in it. Since some of these faces overlap on each other, there are twenty-four resultant faces.

TWO STATES OF UNIVERSE

The three-dimensional world is represented by a cube. There are two cubes in the tesseract. By definition, tesseract represents the bounded transition of a cube in the fourth dimension. So considering each cube represents the state or instance of the universe, there are two instances of universe in tesseract. *Fig*. 5.4 shows the state transition of the universe from state 1 to state 2.

BOUNDED OR CONSTRAINED TRANSITION

Tesseract also indicates a constrained transition of the universe, from one state to the next, in the fourth dimension as the state transition updates the interactions between the particles.

INFINITE TRANSITIONS

Tesseract represents the bounded transition of the universe from one state to the next in the fourth dimension. In universe there are an infinite number of particles and each one of them must go under this transition to reach the next state. So the tesseract represents infinite number of transitions.

CYCLE

Finally, both the states represented by tesseract are part of a cycle that repeats continuously in fourth dimension. So, at any given instant, the universe starts from state 1, transitions to state 2 and completes the cycle. The next cycle starts with state 2 and the cycle goes on continuously.

OBSOLETE STATE

As the cycle of the universe state transition continuously goes on, the previous state of the universe, in *fig*. 5.4 the state 1, will become obsolete or non-existent as the state 2 takes the current state of the universe. Similarly, the state 2 will become obsolete as the universe transitions to the next state.

FOURTH DIMENSION

In this section a new dimension is introduced to completely define the three-dimensional universe, we are living in. The concept can be explained in multiple ways and in this section it is explained, starting with the way the world is perceived.

THREE TIMES

Simple regular activities like walking, talking, human heartbeat, astronomical activities like rotation of the earth take observable amount of time.

Similarly, for the activities that take longer time to complete, it is easy to observe the part of the activity that is finished, a phase of the activity that is progressing and the rest of the activity that is going to happen. In other words, the past, present and the future states of the activity can be observed.

For any event in the three-dimensional space, we can define

the past, present and future states based on the observer time of reference.

The activities that we undertake always span over more than a few seconds, and these activities will only make sense when the whole span is combined.

For example, take a simple word "hello". It takes few seconds to finish saying the word and until the event finishes the sound doesn't make any sense. But still an observer can pick the past–present–future states in the word h-e-l-l-o as it is progressing in the time-line.

BASE SPAN

This way any event can be sub-divided into the smallest possible time and define the past, present and the future time frames. Even though it is difficult for the human brain to observe such a minute time spans, conceptually an event span can be dissected into micro- or even nano-portions.

But then finally we will reach a state, where the event span cannot be sub-divided anymore to observe the past, present and future states of the event span.

This indivisibility of an event can be compared with the Planck's time. Mathematically, it may be possible to write a number way smaller than the Planck's time, but in reality, it is not possible to infinitely divide the event spans occurring in the real world.

This indivisible span is the base span, or building time span of any event span that is observed.

Similarly, the whole event path can be marked in terms of base spans. There will be countless number of these base spans in any given event and they all integrate to make the complete path of the event.

SPAN INTEGRAL

The total time taken for any event is accumulation of all these base spans. The total span *T* of the event can be expressed in the terms of the base span *dt*, the indivisible time span:

$$\int_0^N dt = T$$

where *N* is total number of base spans in the event.

And the total event *E* can be expressed as

$$\int_0^N f(t)dt = E$$

where *f*(*t*) is the a fractional state change in the event *E* in the given base span *dt*.

BASE SPAN AND PRESENT

As explained previously, any event that we observe can be divided into countless number of base spans and the past, present and the future spans. But in reality what really exists? The past, the event portion that is finished does not exist and the remaining portion of the event that is going to come in the future also does not exist. So what really exists is the present. The past and future states are only storage of human memory inside the brain.

All the entities in this universe and all the interactions exist only in the present. If the present is only reality then how small is the present? Because the span of present cannot be null, otherwise there is no presence.

Universe as Fraction

If everything exists only in the present then how fraction is the existence of the universe at any given time?

The base span that is introduced in the previous section is nothing but the present of the universe. So each event is integral of countless number of these presents that come one after the another.

CYCLE

But at the same time, a single span cannot exist in reality. Because having a single span means the entire universe is in a standstill state which is impossible.

A base span that is discussed above cannot exist by itself but it exists in collaboration with the next base span. The state of the continuum represented by the base span is actually in a quasi-equilibrium state that it has to continuously move to the next state, thereby completing the cycle.

So, the universe in a given cycle transitions itself from state 1 to state 2 and updates all the events and the objects.

Fig. 5.5 shows the cyclic nature of the continuum. Each rectangle indicates the schematic representation of the state of the continuum. This cyclic behaviour can be explained very well with tesseract.

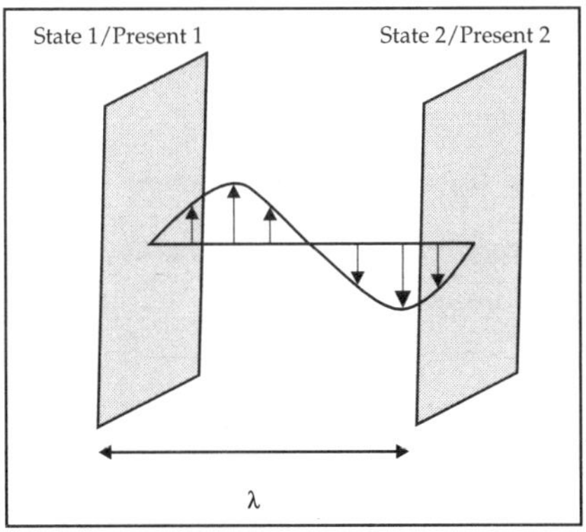

fig. 5.5: Cyclic Universe

As illustrated in the *fig.* 5.6, the first cube on the left-hand side indicates the state of the universe in the current present, which means this state is the starting point, or base for the transition to the next state. The universe in every cycle starts

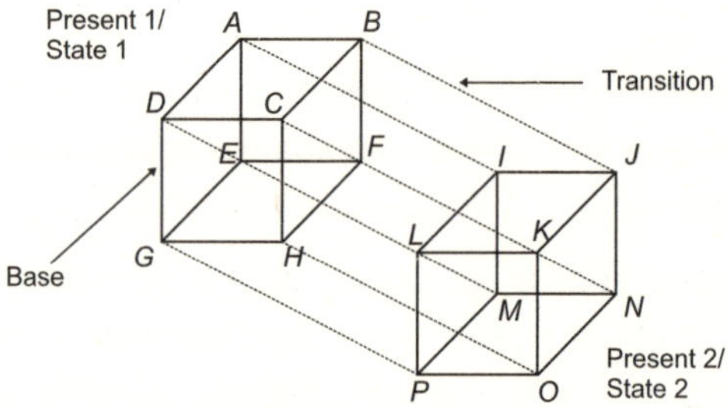

fig. 5.6: Universe representation

with a base state or base present and goes through a constrained transition, controlled by the internal interactions, to the next state or the present.

FREQUENCY

If the span of the continuum, or one cycle of the continuum, is represented by the Planck's time, then how many cycles occur per second, or what is the frequency of this universe?

To estimate the frequency, electromagnetic waves are the best pick as they are no mass carrying and are the fastest moving rays.

Velocity of the light C: 3×10^8 m/sec.

Taking minimum possible distance, as defined by modern science as the wavelength of light λ, i.e. the Planck's length: 1.616×10^{-35} m

The frequency of the continuum:

$$v = \frac{c}{\lambda} = \frac{3 \times 10^8}{1.616 \times 10^{35}} = 1.856 \times 10^{43} \text{ HZ}$$

Assuming Planck's time as the base span of the universe, the v the frequency computed by the above equation is the frequency of the continuum.

This means the universe updates itself 1,044 times per second, creating the illusion of three-dimensional continuously moving universe.

The span of "base span" or the present or the time for which the continuum really exists is 10^{-44} seconds. This value is only hypothetical, it may be impossible to correctly define this number. But just for the convenience, to have a feel of the frequency or the existence of the continuum at any given instance, this span is derived.

DEFINITION

The frequency, ν, discussed above is nothing but the fourth dimension of the three-dimensional spatial universe, providing the smooth and continuous time-line as we see it.

FOUR-DIMENSIONAL UNIVERSE OR CONTINUUM

The activities and interactions of the humans span more than one single base span or the present. Human mind has the capacity to remember the universe for a significant amount of span to perform the activities. Even though the continuum exists only for a fraction of time, human brain makes it far bigger than that. The basic activities we observe like heartbeat, breathing, have time span of the order of multiple of 10^{44} instances of the continuum as most of these basic activities last more than a few seconds.

Fig. 5.7 illustrates the virtual span of the continuum projected onto the human brain. The powerful human nervous system manages to remember multiple spans of the continuum to create a continuous time-line.

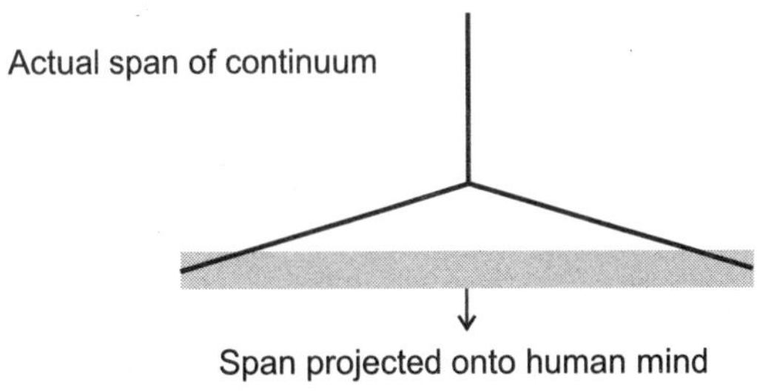

fig. 5.7: Virtual projection

With this new concept of the fourth dimension and the fraction existence of the universe, the continuum properties are explained.

CONTINUUM AS A FLASH

The three-dimensional universe containing the earth, water, air, galaxies, nebulas is so huge in its three-dimensional space that it exists only as a flash, a present, a single frame (a span less than 10^{-44} seconds).

TWO INSTANCES PER CYCLE

In every cycle the universe achieves two distinct states. In every cycle of the continuum, the current present or state is the basis and due to interactions the entities transition to the next state.

ONE SINGLE ENTITY COMPRISING ALL

Even though the entities — both living and non-living — look separated in the three-dimensional universe, in the four-dimensional continuum they all go to the next present at a time, thereby gaining the oneness. There is no meaning for shape and separation (or vacuum) due to this oneness; everything is nothing but just one single entity, the four-dimensional continuum. (This feature can be illustrated with computer screen display. Even though we see various objects on the screen, they all get updated as the screen gets the update. The objects on display may look distinct and behave differently but they are one single entity, in this case the computer screen.)

CYCLIC BEHAVIOUR

The continuum is so huge and exists only for a fraction of time. By transitioning from the current present to the next in a given cycle, continuum manages all the shapes and interactions. With this cyclic behaviour it updates itself countless number (~10^{44}) of times per second to create the virtual three-dimensional world.

WAVE NATURE

If the continuum exists only for such a fraction of time, then is there any time for the matter to really move? It can be derived that there is no time for it, therefore, there is no actual displacement of the particles but everything is nothing but a waveform travelling through the continuum structure, just like the way sound waves travel through the medium without affecting it.

WAVE AND PARTICLE NATURE OF THE CONTINUUM

The particle nature of the continuum can be observed at its base span level. At the same time the continuum shows continuous wave behaviour at the larger time-scales.

If one base span is considered, the whole continuum exists only for that span showing discrete nature or the particle nature. If higher time spans are considered, the continuum exhibits continuous three-dimensional static behaviour, preserving the continuity of the entities and their interactions.

CONTINUUM: A NEW PERSPECTIVE

With these concepts discussed, Table 5.1 highlights the new perspective on the static world humans observe.

The concept of fraction–living continuum is derived based on simple observations. This concept is very powerful because it has a potential to change the way humans deal with the world by proving the static, permanent world as fraction–living continuum.

VIRTUAL CONTINUUM

As explained, continuum, even though such a vast three-dimensional entity, exists only for a fraction of time at any given instance.

Table 5.1: The New Perspective on the Static World Humans Observe

No.	*Feature*
1	Oneness — single entity due to the fourth dimension
2	Three-dimensional world (visible) and fourth dimension (underneath)
3	Fraction time existence — refreshes continuously creating virtual 3-D world
4	Cyclic — transitions from current state/present to the next state in a constrained way — tesseract representation
5	Wave nature — pure energy, virtual particle

Simple words like "hello" or basic actions like human breathing take more than trillions of presents. Since there is only one present existing at any given time, these actions or entities do not exist as they are. They are only an accumulation effect of space and the span of the continuum.

Just like the movies are virtual, at any given time only one frame of the movie exists but not all the movie, all the activities and entities we observe are virtual. They are virtual because they do not exist as they are but they are just a projection on to the human memory.

MANIFESTED VIRTUAL UNIVERSE

With the above discussion, all the entities and activities that we observe can be considered as virtual. In other words, the manifested universe is completely virtual in nature.

NON-MANIFESTED REALITY

In such a fraction existing continuum, what really exists? It can be logically illustrated that the non-manifested structure of the continuum is the one that really exists in such a infinitesimal span. Just like the way the dark energy and dark matter are completely transparent, the structure of the continuum can be considered as non-manifested entity or reality.

This concept is the basis of all the concepts discussed in this report and it is the centre of consciousness normalization.

Consciousness and Host

The previous section discussed about the true nature of the field, the continuum. This section will summarize the true nature of the observer by analysing the relationship between the consciousness and the human body. The detailed analysis can be found in the books referred (Jata 2011, 2012).

CONSCIOUSNESS AND THE HOST

Consciousness and its host, the body, are considered as separate but not independently exclusive entities. In other words, consciousness cannot be identified with the host and what the host does, but at the same time, consciousness cannot exist without an active host.

fig. 5.8: Consciousness

In simple terms, *fig*. 5.8 illustrates the consciousness and its host. The host is shaped irregularly to highlight that the host can be a single cell organism or a human body. The consciousness is indicated as a small area just for illustration purpose, but it can be assumed that consciousness prevails across the host.

HOMOEOSTASIS

Along with autonomic nervous system, endocrine system, hypothalamus that connects the both, the liver and kidneys maintain the internal stability of the human body. These systems respond, based on the changes in external environment and internal take into body. Homoeostasis state, for the most part, is managed by human body without any help from consciousness or without any awareness that the mechanism is in place.

This homoeostasis balancing is in place across multiple systems in the human body. Nervous system, endocrine system, skeletal system and various other systems are under constant regulation to maintain the equilibrium state of the host in an involuntary way, without any initiation from the consciousness.

INDEPENDENT HOST

Most of the major functions of the human body are managed either completely below consciousness or partially below the consciousness level. Body can inform the consciousness on as needed basis, so that it can use CNS to make the decisions.

In a normal, relaxed environment, human body can operate without any initiation from the consciousness. As shown in the *fig*. 5.9, host can interact with consciousness as and when needed.

Unless there is some requirement from the host, consciousness can ignore its host.

PERFECT ENGINE

The connection between the consciousness and its host can be explained very well using modern car and the driver

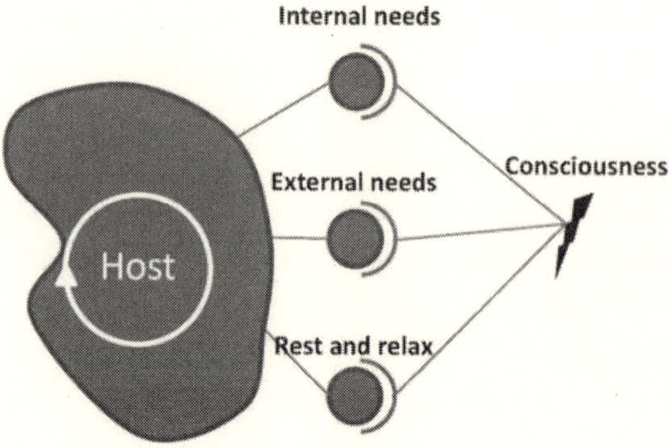

fig. 5.9: Host interactions

interactions. While cruising on a freeway without traffic, and assuming the car is in perfect condition, the driver doesn't have to meddle with the state of the car. Only when there is some traffic on the road, or if car needs some gas, the driver has to involve himself and take care of the situation. The driver can ignore the car and the road but at the same time be ready to act on as needed basis. Similarly, human body, as a perfectly balanced engine, does not need the interference from the consciousness unless there is some internal need or there is some external environmental situation.

The consciousness as the driver of the host can completely stay idle and doesn't meddle with the state of the host. More explained about this relationship in the normalization book (Jata 2011).

THREE STATES OF THE CONSCIOUSNESS

Using the schematic diagrams, the three states of the consciousness as explained by Sanātana Dharma can be illustrated as follows:

Waking state

In the first state, due to external awareness, the consciousness associates itself with the worldly objects and activities.

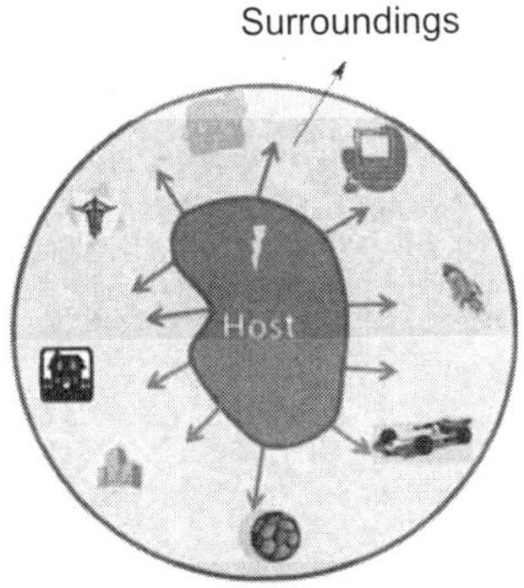

fig. 5.10: Waking state

The arrows pointing out of the host indicate the attachment of consciousness to its surroundings.

Dream State

In the second state, due to the internal awareness, the con-sciousness dreams/thinks about personal feelings.

The arrows directing internal indicate consciousness attachment to the host and activities like dreams and emotions due to the memories of activities during the waking state.

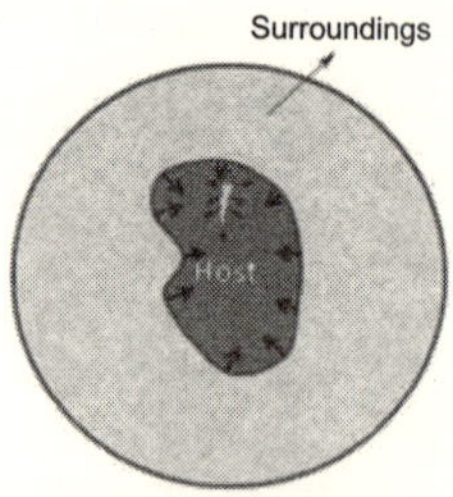

fig. 5.11: Dream state

Deep sleep State

The third state, deep sleep state, is the state in which there are no dealings with the external environment, no internal desires/day/night dreams. The colour of the host and consciousness is shown as dark to indicate that the both are non-responsive to the outside world and inside thinking.

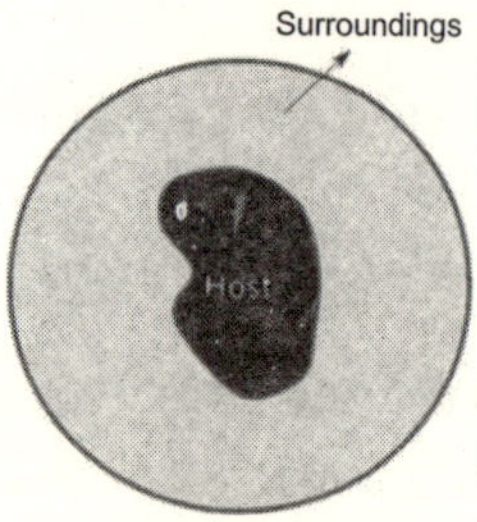

fig. 5.12: Deep sleep state

HOST MECHANISMS

More details are presented in the normalization book (Jata 2011) explaining the human nervous system, the brodmann maps and other concepts, to illustrate the independent nature of the host.

New Interpretations

PURUṢA-SŪKTA

Puruṣa-Sūkta (Chanana 1983) is the centre point of the new interpretations. The model of four-dimensional continuum fits very well with Puruṣa-Sūkta. It appears *dharma* used *puruṣa*, a human model, to explain the four-dimensional universe to a common person.

Using the concept of four-dimensional continuum, a few important *śloka*s from the *sūkta* are interpreted. The concept fits in so well that all the *śloka*s can be interpreted meaningfully in a single context. Not only this *sūkta*, as we see in the following sections, but the major scriptures of the *dharma* also fit very well in one single context, the four-dimensional continuum. An in-depth analysis of various scriptures is presented in the book (Jata 2012).

SINGLE ENTITY

> सहस्रशीर्षा पुरुषः सहस्राक्षः सहस्रपात्।
> स भूमिं विश्वतो वृत्वात्यतिष्ठद्दशाङ्गुलम्॥1॥
>
> *sahasraśīrṣā puruṣaḥ sahasrākṣaḥ sahsrapāt* ।
> *sa bhūmiṁ viśvato vṛtvātyatiṣṭhaddaśāṅgulam* ॥1॥
> — *ṚV* X.90.1
>
> The *Puruṣa* has an infinite number of heads, eyes and bodies. As a single entity the *Puruṣa* is expanded all over the earth, the space and unlike a living body it is beyond the ten fingers of the body.

This is the first *śloka* of the hymn. This *śloka* explains the overall nature of the continuum. This *śloka* emphasizes the oneness nature of the continuum.

In Sanskrit the meaning for सहस्र (*sahasra*) is thousand and also infinite or unlimited. The first two *pada*s of the *śloka* describe that the *Puruṣa* or the continuum has infinite heads (सहस्रशीर्ष *sahasraśīrṣa*), infinite eyes (सहस्राक्षः *sahasrākṣaḥ*) and infinite feet (सहस्रपात् *sahasrapāt*).

This *śloka* greatly emphasizes the oneness of the continuum. Since all the beings are one in the continuum, the *śloka* says the continuum has heads, eyes and legs. By indirectly rejecting that individual beings have bodies, the *śloka* emphasizes that it is not the living beings but it is the continuum/*Puruṣa* who owns the bodies as one single being, and therefore, has infinite heads, eyes and feet.

In order to avoid the confusion that *śloka* is explaining about a real person with so many heads, the third and the fourth *padas* of the *śloka* explain that this *Puruṣa*, even though it has all the living entity parts, it is expanded (वृत्व *vr̥tva*) to the whole earth (भूमिं *bhūmiṁ*) and the universe (विश्वतो *viśvato*) too, therefore it exists beyond (अत्यतिष्ठत् *atyatiṣṭhat*) the physical ten fingers (दशाङ्गुलम् *daśāṅgulam*).

Since *Paramātmā* is nothing but the continuum, it is the continuum that is expanded everywhere.

So the *śloka* fits well with the oneness nature of the continuum:

> As a single entity the continuum is expanded all over the space and all the living entities are part of it and it is the continuum that owns everything we see.

DESIGN

In the first *śloka* above the *sūkta* explains the basic oneness nature of the continuum. In the third *śloka* for the first time the *sūkta* explains what this continuum consists of.

एतावानस्य महिमातो ज्यायाँश्च पूरुषः।
पादोऽस्य विश्वा भूतानि त्रिपादस्यामृतं दिवि॥3॥

etāvānasya mahimāto jyāyāṁśca pūruṣaḥ।
pādo 'sya viśvā bhūtāni tripādasyāmr̥taṁ divi ॥3॥
— *R̥V* X.90.3

All these forms and everything is nothing but *Puruṣa*. Moreover, the *Puruṣa* itself is a masterpiece, a great design. One part consists of all the active elements and remaining three parts hold the infinite, invariant space.

> पादोऽस्य (*pādo 'sya*) = one part (dimension) has विश्वा भूतानि (*viśvā bhūtāni*) = all the moving and त्रिपादस्य (*tripādasya*) = three parts (three dimensions) has अमृतं दिवि (*amṛtaṁ divi*) = the infinite universe.

One part of it has all the moving or transitioning/changing properties and the remaining three parts have the infinite immutable universe.

This *śloka* explains how the continuum is made up of. It highlights the one part (पाद *pāda*), and the remaining three parts (त्रिपाद *tripāda*). The *śloka* explains the three-dimensional part is an endless, immutable space (अमृतं दिवि *amṛtaṁ divi*) and due to the fourth dimension everything moves (विश्वा भूतानि *viśvā bhūtāni*) or transitions to next state. (Again looking at the computer screen example — The screen is the container or space where all the display happens. But it is due to the refresh rate of the screen that the interactions between the objects are made possible. Similarly the three-dimensional universe is an infinite space where everything resides, but it is the fourth dimension that makes the interactions possible.)

Also here the *śloka* did not say two and two but one (पाद *pāda*) and three (त्रिपाद *tripāda*) indicating that the continuum is a four-dimensional universe with one-dimension and the remaining three dimensions together. This *śloka* in essence explains the four-dimensional arrangement of the universe.

Combining the two *śloka*s:

> As a single entity the continuum is expanded all over the space and all the living entities are part of it and it is the continuum that owns everything we see.
>
> All the living and non-living entities are just various forms of the continuum, moreover the continuum itself is magnificent. Three parts of it are spatial dimensions where this infinite, immutable space exists and the fourth part is the one which brings the universe into existence.

BEHAVIOUR

> त्रिपादूर्ध्व उदैत्पुरुषः पादोऽस्येहाभवत्पुनः।
> ततो विष्वङ् व्यक्रामत्साशनानशने अभि॥4॥
>
> *tripādūrdhva udaitpuruṣaḥ pādo 'syehābhavat punaḥ*।
> *tato viṣvaṅ vyakrāmatsāśanānaśane abhi* ॥4॥
>
> — *Ṛ̥V* X.90.4

> Three parts are above the *Puruṣa* and the fourth one repeats continuously. From this arrangement the universe evolved, the forces and the followers.

This *śloka* goes a bit more detailed way, explaining how the constituents of the continuum behave. The three dimensions (त्रिपाद् *tripāda*) of the continuum are situated above (ऊर्ध्व *ūrdhva*), are visible (उदैत् *udait*) to us and the fourth dimension (पादोऽस्येहा *pādo 'syehā*) is the one that refreshes or repeats (भवत्पुनः *bhavat-punaḥ*) and takes the three-dimensional world from one present to the next.

From this the whole universe is evolved.

Combining the three *ślokas*:

> As a single entity the continuum is expanded all over the space and all the living entities are part of it and it is the continuum that owns everything we see. All the living and non-living entities are just various forms of continuum, moreover the continuum itself is magnificent. Three parts of it are spatial dimensions where this infinite, immutable space exists and the fourth part is the one which brings the universe into existence. These three parts or dimensions, where the infinite space resides, are visible to the eye. The fourth part of the dimension is the one that refreshes/repeats continuously bringing the remaining three parts into existence. From this arrangement the whole universe is evolved.

ENGINE

सप्तास्यासन् परिधयस्त्रिः सप्त समिधः कृताः।
देवा यद्यज्ञं तन्वाना अबध्नन् पुरुषं पशुम्।।15।।

saptāsyāsan paridhayastriḥ sapta samidhaḥ kṛtāḥ ।
devā yadyajñaṁ tanvānā abadhnan puruṣaṁ paśum ।।5।।

— *ṚV* X.90.15

In this *Puruṣa* arrangement, one set (1 x 7) is the boundary and the remaining three sets (3 x 7) is the base. The nobles executed this dedicated act by making the *Puruṣa* as sacrifice.

This *śloka* explains how the universe is running like an engine for the past billions of years. This *śloka* fits well with the continuum transition as explained by the tesseract model.

Fig. 5.13 illustrates the *śloka*. This *śloka* highlights the major concepts of modern physics: Since mass is energy and light is electromagnetic radiation/energy, the visible mass is nothing but electromagnetic energy.

So in this *śloka* by using "seven" for the light spectrum or energy, the *sūkta* appears to explain that the three-dimensional world (त्रि:सप्त *triḥsapta*) itself is energy and the fourth dimension

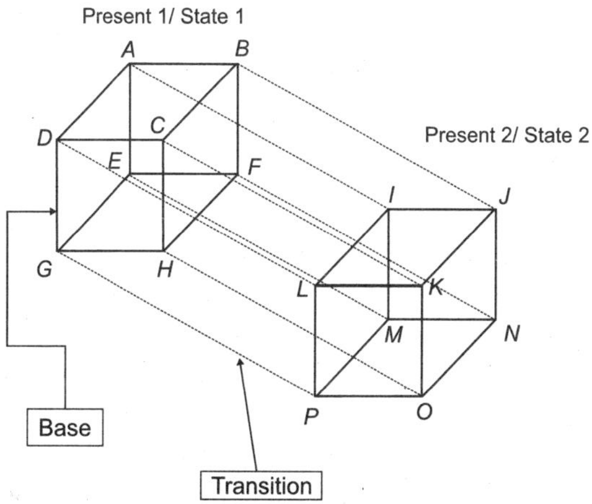

fig. 5.13: Cycle by tesseract

(सप्त *sapta*) or the transition to the next state is also energy, indicating the whole four-dimensional continuum is an energy signature that exists like a lightning in any given present.

Fig. 5.13 shows one cycle of the continuum. So the *śloka* says the continuum starts with the left hand side cube, the three-dimensional universe representation (त्रिःसप्त *triḥsapta*) at the current present, as a basis or starting state of the cycle (समिधः कृताः *samidhaḥ kṛtāḥ*) and transitions to the next state in the fourth dimension (सप्तासयासन् *saptāsayāsan*), along the boundary shown by the dashed lines (परिधयः *paridhayaḥ*). The *śloka* indicates the transition is very constrained, controlled by various forces or interactions between entities in a given state.

देवा यद्यज्ञं तन्वाना अबध्नन् पुरुषं पशुम्।।

devā yadyajñaṁ tanvānā abadhnan puruṣaṁ paśum ।।

देवा (*devā*) = nobles, यत् (*yat*) = this, यज्ञं (*yajñaṁ*) = dedicated act, तन्वाना (*tanvānā*) = performed, अबध्नन् (*abadhnan*) = binding, पुरुषं (*puruṣaṁ*) = the continuum, पशुम् (*paśum*) = as sacrifice

This way, this cyclic universe that has been going on since the beginning of times, is performed by sacrificing the previous state of the continuum in every given cycle. The complete translations to *sūkta* are provided in the *dharma* interpretation book (Jata 2012).

Combining the four *śloka*s:

> As a single entity the continuum is expanded all over the space and all the living entities are part of it and it is the continuum that owns everything we see. All the living and non-living entities are just various forms of the continuum, moreover the continuum itself is magnificent. Three parts of it are spatial dimensions where this infinite, immutable space exists and the fourth part is the one which brings the universe into existence. These three parts or dimensions where the infinite space resides, is visible to the eye. The

Table 5.2: Puruṣa-Sūkta — Continuum

Śloka	*Title*	*Description*
1	Definition	Overall structure — oneness
2	Design	Dimensions — three and one
4	Behaviour	Three-dimensional world and refresh
15	Engine	Transition and cycle

fourth part or the dimension is the one that refreshes/repeats continuously brings the remaining three parts into existence. From this arrangement the whole universe is evolved. **Every time this three-dimensional space, as a base, transitions to the next state in a constrained way. The nobles participated in this continuous act by sacrificing the previous instance of the continuum.**

So far most crucial *śloka*s are selected to support *Paramātmā* as four-dimensional continuum. Summarizing the progression of *śloka*s *sūkta* followed to explain the four-dimensional continuum:

TESSERACT IN ŚVETĀŚVATARA UPANIṢAD

Puruṣa-Sūkta fits very well with *Paramātmā* as the four-dimensional continuum model. The Sanātana Dharma also provided the mathematical formulation for *Para-Brahman* or four-dimensional continuum, the tesseract, in *śloka* 4 of *Śvetāśvatara Upaniṣad* (Tyagisananda 1986).

तमेकनेमिं त्रिवृतं षोडशान्तं शतार्धारं विंशतिप्रत्यराभिः।
अष्टकैः षड्भिर्विश्वरूपैकपाशं त्रिमार्गभेदं द्विनिमित्तैकमोहम्।।4।।

tamekanemiṁ trivṛtaṁ ṣoḍaśāntaṁ
śatārdhāraṁ viṁśatirpatyarābhiḥ ।
aṣṭakaiḥ ṣaḍbhirviśvarūpaikapāśaṁ
trimārgabhedaṁ dvinimittaikamoham ।।4।।

This *Puruṣa, Paramātmā* has one and three cyclic arrangement with sixteen ends. There are fifty connections with each of them having twenty sub-connections. There are eight entities with six sub-entities in each of them. There is a belt of universal size, arranged in three ways, and the *Puruṣa* becomes two in every cycle.

Summarizing the tesseract properties highlighted by the *śloka* (Table 5.3):

Table 5.3: Tesseract in Upaniṣads

Tesseract Property	*Śloka Description*
4-D Representation	*ekanemiṁ, trivr̥taṁ*
Starts with two cubes	*ṣaḍaśāntaṁ*
Connect the cubes to make one entity	*viśvarūpaikapāśaṁ*
Results in total eight cubes	*aṣṭakaiḥ, ṣaḍbhiḥ*
Represents two states of universe	*dvi + nimitta*
Represents one cycle of 4-D space	*aikamoham*
Infinite transitions	*śatārdhāraṁ viṁśatirpatyarābhiḥ*

The features of tesseract fit well with *śloka* description. By illustrating the mathematical representation of four-dimensional space, the tesseract, *Śvetāśvatara Upaniṣad* completes the definition of *Paramātmā*.

The next two sections interpret some intriguing *śloka*s from the Upaniṣads and the *Bhagavad-Gītā* to prove that with this new interpretation, these *śloka*s really make sense and the translations perfectly mend with the context. More detailed translations are provided in the *dharma* interpretation book (Jata 2012).

Upaniṣads

MUṆḌAKOPANIṢAD

ब्रह्मैवेदममृतं पुरस्ताद् ब्रह्म पश्चाद् ब्रह्म दक्षिणतश्चोत्तरेण।
अधश्चोर्ध्वं च प्रसृतं ब्रह्मैवेदं विश्वमिदं वरिष्ठम् ॥2.2.11॥

brahmaivedamamr̥taṁ purastād brahma paścād brahma dakṣiṇataścottareṇa ।

adhaścordhva ca prasṛtaṁ brahmaivedaṁ viśvamidaṁ
variṣṭham ||2.2.11||

Paramātmā is the imperishable. *Puruṣa* is in the front, in the back, to the right, to the left, below above spread out in this whole three-dimensional universe.

Puruṣa-Sūkta explains the four-dimensional continuum as *Paramātmā*. Even though the entities have a life span in three-dimensional universe, the four-dimensional continuum as a single entity will never perish. It is the continuum that is in the front–back, right–left, top–bottom or the x, y and z directions of the three-dimensional space. So the whole universe is nothing but this great *Paramātmā*.

यदर्चिमद्यदणुभ्योऽणु च यस्मिँल्लोका निहता लोकिनश्च।
तदेतक्षरं ब्रह्म स प्राणस्तदु वाङ्मनः तदेतत्सत्यं तदमृतं
तद्वेद्धव्यं सोम्य विद्धि।।2.2.2।।

yadarcimadyadaṇubhyo 'ṇu ca yasmiṁllokā
nihatā lokinaśca|
tadetakṣaraṁ brahma sa prāṇastadu vāṅmanaḥ
tadetatsatyaṁ tadamṛtaṁ tadveddhavyaṁ
somya viddhi ||2.2.2||

Shining brilliantly, *Paramātmā* exists as a fraction of a fraction. In this all the worlds are resided. Also the living beings exist in this fraction. He is imperishable, real, permanent, knowledge; the realized ones know this.

As explained in the previous sections, the continuum exists like a flash/refresh at any given time. The *śloka* explains the *Paramātmā* is shining or brilliant (अर्चिमत् *arcimat*) like a fraction of fraction (अणुभ्यः अणु *aṇubhyaḥ aṇu*) to represent the main feature of the continuum. Inside this brilliant continuum (यस्मिँ *yasmiṁ*), the worlds (लोकाः *lokāḥ*) are established/situated (निहता *nihatā*) and the living beings too (लोकिनः *lokinaḥ*).

The third and fourth *pada*s explain *Paramātmā* is explained by the realized ones (सोम्य *somya*) as everything: immutable (अक्षरं, अमृतं *akṣaraṁ, amṛtaṁ*), the living forces and interactions (प्राणः, वाङ्मनः *prāṇaḥ, vāṅmanaḥ*) and knowledge (वेद्धव्यं *veddhavyaṁ*).

वृहच्च तद् दिव्यमचिन्त्यरूपं सूक्ष्माच्च तत् सूक्ष्मतरं विभाति।
दूरात् सुदूरे तदिहान्तिके च पश्यन्त्विहैव निहितं गृहायाम्।।3.1.7।।

vṛhacca tad divyamacintyarūpaṁ sūkṣmācca tat sūkṣmataraṁ vibhāti।
dūrāt sudūre tadihāntike ca paśyantivihaiva nihitaṁ gṛhāyām ।।3.1.7।।

Being fractional, this *Puruṣa* is brilliant, great and beyond imagination. The *Puruṣa* is there faraway and it is here too. The realized ones are merged with this greatness.

Great (बृहत् *bṛhat*) is this brilliant (दिव्यम् *divyam*) structure which is beyond imagination (अचिन्त्यरूपं *acintyarūpaṁ*) and is very minute or with fractional existence (सूक्ष्माच्च *sūkṣmācca*). In this (तत् *tat*) fractional behaviour (सूक्ष्मतरं *sūkṣmataraṁ*) manifests brilliantly (विभाति *vibhāti*). Also he is far away (दूरात् सुदूरे *dūrāt sudūre*) and inside (अन्तिके *antike*) here too. The realized ones reside (निहितं *nihitaṁ*) with this great oneness (गृहायाम् *gṛhāyām*).

KAṬHA UPANIṢAD

अग्निर्यथैको भुवनं प्रविष्टो रूपं रूपं प्रतिरूपो बभूव।
एकस्तथा सर्वभूतान्तरात्मा रूपं रूपं प्रतिरूपो बहिश्च।।5.9।।

agniryathaiko bhuvanaṁ praviṣṭo rūpaṁ rūpaṁ pratirūpo babhūva।
ekastathā sarvabhūtāntarātmā rūpaṁ rūpaṁ pratirūpo bahiśca।।5.9।।

Just like the fire takes the shape of the entity it assimilates, *Paramātmā*/continuum is inside all the entities, shapes and it is outside them too.

वायुर्यथैकोभुवनं प्रविष्टो रूपं रूपं प्रतिरूपां बभूव।
एकस्तथा सर्वभूतान्तात्मा रूपं रूपं प्रतिरूपो बहिश्व॥5.9॥

vāyuryathaikobhuvanaṁ praviṣṭo rūpaṁ rūpaṁ pratirūpāṁ babhūva।
ekastathā sarvabhūtāntātmā rūpaṁ rūpaṁ pratirūpo bahiśva॥5.9॥

Even though one, as the air takes the shape of the entity it enters, *Paramātmā*/continuum is inside all the entities, shapes and it is outside of them too.

बालाग्रशतभागस्य शतधा कल्पितस्य च।
भागो जीवः स विज्ञेयः स चानन्त्याय कल्पते॥5.9॥

bālāgraśatabhāgasya śatadhā kalpitasya ca।
bhāgo jīvaḥ sa vijñeyaḥ sa cānantyāya kalpate॥5.9॥

Hundredth part of the hair, hundredth part, like this situated and also along with this minuteness, need to realized that this *ātmā* is situated as infinite, thus made possible.

This *śloka* highlights the four-dimensional nature of the continuum. Even though the universe is so vast, it exists only for a fraction of time, a time span that represents the cycle of the continuum.

The *śloka* highlights this nature, stating that *Paramātmā* existence is so minute, fraction part of hair width to represent the time span of the four-dimensional continuum. At the same time it describes his existence as infinite — the size of the continuum in the three-dimensional space.

सूक्ष्मातिसूक्ष्मं कलिलस्य मध्ये विश्वस्य स्रष्ठारमनेकरूपम्।
विश्वस्यैकं परिवेष्टितारं ज्ञात्वा शिवं शान्तिमत्यन्तमेति॥4.14॥

sūkṣmātisūkṣmaṁ kalilasya madhye viśvasya srasṭhāramanekarūpam।
viśvasyaikaṁ pariveṣṭitāraṁ jñātvā śivaṁ śāntimatyantameti॥4.14॥

> In between the fractional existing solid heap, the entire universe entities exist as one. Realizing this, the nobles attained peace by joining the one.

Even though there are different shapes in the three-dimensional world, they achieve oneness in the four-dimensional continuum and exist only for a fraction time. Assuming the existence of a base energy signature in such a fraction time, the whole universe is made up of infinite number of these minute energy particles.

The *śloka* explains the universe is a solid heap (कलिलस्य *kalilasya*) of these minute particles (सूक्ष्मातिसूक्ष्मं *sūkṣmātisūkṣmaṁ*) and in between them (मध्ये *madhye*) are the various shapes (अनेकरूपम् *anekarūpam*) of *Paramātmā* (स्रष्ठारम् *srasṭhāram*). This variety is enveloped (परिवेष्टितारं *parivesṭitāraṁ*) by the oneness (एकं *ekaṁ*) of the continuum, knowing this the realized ones (शिवं *śivaṁ*) calmly (शान्तिम् *śāntim*), unconditionally attain (एति *eti*) the *Paramātmā* (अत्यन्तमं *atyantamaṁ*) or the oneness, instead of identifying themselves with the body.

KENOPANIṢAD

> यद्वाचाऽनभ्युदितं येन वागभ्युद्यते॥1.5.1॥
> यन्मनसा न मनुते येनाहुर्मनो मतम् ॥1.6.1॥
> यच्चक्षुषा न पश्यति येन चक्षूँषि पश्यति॥1.7.1॥
> यच्छ्रोत्रेण न शृणोति येन श्रोत्रमिदं श्रुतम् ॥1.8.1॥
> यत्प्रणेन न प्राणिति येन प्राणः प्रणीयते॥1.9.1॥
> तदेव ब्रह्मा त्वं विद्धि नेदं यदिदमुपासते॥1.9.2॥

> *yadvācā 'nabhyuditaṁ yena vāgabhyudyate*॥1.5.1॥
> *yanmanasā na manute yenāhurmano matam*॥1.6.1॥
> *yaccakṣuṣā na paśyati yena cakṣūm̐ṣi paśyati*॥1.7.1॥
> *yarcchotreṇa na śṛṇoti yena śrotramidaṁ śrutam*॥1.8.1॥
> *yatpraṇena na prāṇiti yena prāṇaḥ praṇīyate*॥1.9.1॥
> *tadeva brahmā tvaṁ viddhi nedaṁ yadidamupāsate*॥1.9.2॥

> That by speech cannot be expressed and by which speech is made possible. That by sense cannot be expressed and by which mind is made possible. That by eyes cannot be seen and by which vision is made possible. That by ears cannot be heard and by which ears are made possible. That is how the *Paramātmā* is situated, but not the one that is worshipped here.

The above-mentioned are the first two *pada*s of four *śloka*s (marked as 5.1, 6.1, 7.1, 8.1 and 9.1) and each *śloka* has the same third and fourth *pada*s (marked as ।।1।।).

The continuum is responsible for everything in the universe, even though it exists for only a fraction of time. The *śloka*s say even though the continuum is responsible for the creation of the sensory organs for the hearing, talking, seeing or the consciousness, none of these is useful to observe the four-dimensional nature of the continuum. The *śloka*s further explain that the continuum or the *Paramātmā* is not the gods that the people are praying and/or worshipping.

Bhagavad-Gītā

CONTINUUM

> प्रकृतिं स्वामवष्टभ्य विसृजामि पुनः पुनः।
> भूतग्राममिमं कृत्स्नमवशं प्रकृतेर्वशात्।।9.8।।
>
> *prakṛtiṁ svāmavaṣṭabhya visṛjāmi punaḥ punaḥ* ।
> *bhūtagrāmamimaṁ kṛtsnamavaśaṁ prakṛtervaśāt* ।।9.8।।
>
> प्रकृतिं (*prakṛtiṁ*) = the world, स्वाम अवष्टभ्य (*svāma avaṣṭabhya*) = rested inside, *Paramātmā* विसृजामि (*visṛjāmi*) = flow out/created, पुनः पुनः (*punaḥ punaḥ*) = again and again, भूतग्राम (*bhūtagrāma*) = all the entities, इममं (*imamaṁ*) = this, कृत्स्नम् (*kṛtsnam*) = all, अवशं (*avaśaṁ*) = automatically, प्रकृते (*prakṛte*) = inside interactions, वशात् (*vaśāt*) = controlled by
>
> Being part of *Paramātmā*, all the entities in the universe are ebbed out again and again.

This *śloka* explains the cyclic nature of the continuum. The three-dimensional world (प्रकृतिं *prakṛtiṁ*) as a part of four-dimensional continuum (स्वाम अवष्टभ्य *svāma avaṣṭabhya*) ebbs out or creates (विसृजामि *visṛjāmi*) again and again (पुनः पुनः *punaḥ punaḥ*) in every cycle or present. Here the *śloka* uses the term "flows out/ebbs out" as current present evolves out of the previous one. During this transition, all the entities (भूतग्राममिमं *bhūta-grāmamimaṁ*) in the world automatically enter (अवशं *avaśaṁ*) to the next controlled state due to the resultant interactions (प्रकृतेः वशात् *prakṛteḥ vaśāt*) between them.

तत्रैकस्थं जगत्कृत्स्नं प्रविभक्तमनेकधा।
अपश्यद्देवदेवस्य शरीरे पाण्डवस्तदा।।11.13।।

tatraikasthaṁ jagatkṛtsnaṁ pravibhaktamanekadhā ।
apaśyaddevadevasya śarīre pāṇḍavastadā ।।11.13।।

Even though the entities in universe are divided into so many, they are all one single entity, thus sees Arjuna.

This *śloka* describes the oneness nature of the four-dimensional continuum. Even though the entities in three-dimensional world look separated or divided (प्रविभक्तम *pravibhaktam*) into specific shapes (अनेकधा *anekadhā*), they all are (जगत्कृत्स्नं *jagatkṛtsnaṁ*) nothing but one single entity (एकस्थं *ekasthaṁ*), this four-dimensional continuum.

In the three-dimensional world all the living entities and non-living entities (सर्वाणि भूतानि *sarvāṇi bhūtāni*) change (क्षरः *kṣaraḥ*) continuously. These entities attain oneness (कूटस्थः *kūṭasthaḥ*) in the four-dimensional continuum and it is the one that is permanent and never changes (अक्षरः *akṣaraḥ*).

कविं पुराणमनुशासितारमणोरणीयंसमनुस्मरेद्यः।
सर्वस्य धातारमचिन्त्यरूपमादित्यवर्णं तमसः परस्तात्।।8.9।।

kaviṁ purāṇamanuśāsitāramaṇoraṇīyaṁsamanusmaredyaḥ ।
sarvasya dhātāramacintyarūpamādityavarṇaṁ tamasaḥ parastāt ।।8.9।।

> Beginning of everything, controller of everything, situated like a fraction, repeats itself. Maintainer of everything, beyond imagination, pure energy, dark to our senses and infinite.

The continuum is beginning of everything (कविं पुराणम् *kaviṁ purāṇam*) and controller of everything (अनुशासितारम् *anuśāsitāram*). Even though looks so big in three-dimensional world it exists only for a fraction of time (अणोरणीयं *aṇoraṇīyaṁ*) which is like a smallest particle. It repeats itself (समनुस्मरेद्य: *samanusamredyaḥ*) as it moves to the next present taking the previous state into consideration. It is everywhere (सर्वस्य *sarvasya*), maintainer of everything (धातारम् *dhātāram*), beyond imagination (अचिन्त्यरूपम् *acintyarūpam*). It is pure energy (आदित्यवर्णं *ādityavarṇa*), dark to our senses and infinite (तमस: परस्तात् *tamasaḥ parastāt*).

Consciousness Concepts

Previous section presented the independent nature of the human body and explained its relation with the consciousness. Similarly, *dharma* also differentiates the host and the consciousness, and introduces concepts like *ātmā* (soul) to highlight the separation. These *śloka*s are referred from the *Bhagavad-Gītā* (Prabhupada 1997).

> The bewildered spirit soul, under the influence of the three modes of material nature, thinks him to be the doer of activities, which are in actuality carried out by the nature.
> — III.27

> As the Sun alone illuminates this entire universe, so does the living entity, one within the body, illuminate the entire body by consciousness. — XIII.34

> The sky, due to its subtle nature, does not mix with anything, although it is all-pervading. Similarly, the soul does not mix with the body though situated in that body. — XIII.33

As we will see in the next section, there is a clear differentiation between the consciousness and the host which can be explained, based on the virtual nature of the continuum or *Paramātmā*.

Normalization

This section integrates the field and the observer to introduce a new way of living called "normalization of consciousness" i.e. this part of the report defines guidelines for the normalized behaviour of consciousness with respect to its host and the nature of the outside world. More can be found in the normalization book (Jata 2011).

THE FIELD

Recapping the basic characteristics of the continuum discussed so far:

- The three-dimensional world appears continuous and permanent.
- As a four-dimensional continuum, it exists only for a fraction of time at any given instant.
- Even the entities, both living and non-living, appear distinct and separate in the three-dimensional universe, they lose their identity and become part of the four-dimensional continuum.
- Matter is nothing but combination of elementary particles, which are nothing but standing waves moving through the continuum structure.
- In reality, the continuum structure is the only one that really exists and it exists in a non-manifested way.
- All the entities and activities that span more than one present are virtual as only one present exists at any given time.
- Irrespective of human perspective, everything in the continuum exists for only a fraction of time and there is nothing else beyond this span.

THE OBSERVER

Summarizing the characteristics of the host and the consciousness:

- Consciousness is the basic quality that differentiates living from non-living and it is the common feature among all the living beings.
- Consciousness is not defined by what host can perform.
- Consciousness and its host are two separate and dependent entities.
- Host by itself is very sophisticated system and for the most part it performs the activities "below consciousness" and it can inform the consciousness when needed.

INTEGRATING HOST AND THE CONTINUUM

Using the characteristics of the field, the four-dimensional continuum, the nature of the consciousness host can be redefined.

HUMAN BRAIN

Even though the field or the continuum exists for a fraction of time, the human observer's central nervous system has the ability to store the multiple spans of the continuum to create the illusion of a static world that exists beyond the reality.

Human brain combines the past, present and future of the activities of an event to give the illusion of the continuous span far bigger than the span of the continuum.

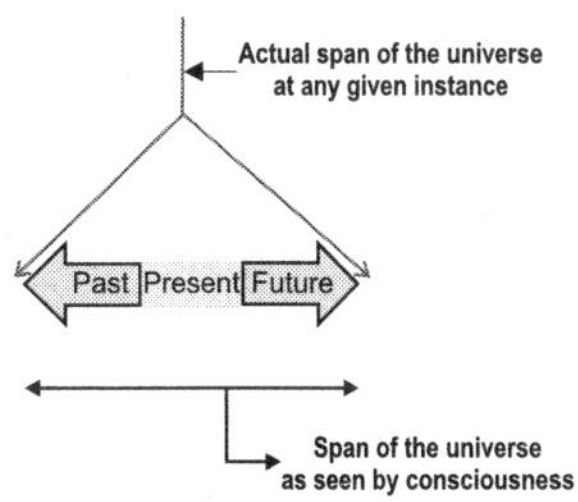

fig. 5.14: Human world

The basic activities performed by humans span over more than one present. For example, the span of heartbeat is about 0.833 second which is far bigger than the span of the continuum. So, in terms of number of instances of the continuum, the span of the heartbeat is:

$$\frac{0.833}{10^{-44}} = 8.33 \times 10^{43}.$$

About 1,044 instances of continuum pass each heartbeat.

This way, human brain facilitates the mechanism for the consciousness to perform activities in the world. On the other side, due to the same mechanism, consciousness perceives the universe more than what it is — as a static, permanent world.

VIRTUAL HOST

Considering all the entities and activities we observe are virtual, it can also be explained that the host itself, being part of virtual field, is virtual.

By definition, consciousness or free will is not under the influence of laws of nature. In other words, as the state of non-living entities is defined by the forces of interactions among them, the free will has the luxury of controlling its course of action.

As the free will is beyond the forces of nature, being virtual by nature, the human body must follow the laws of nature.

The host by itself does not have any intelligence to make decisions. Even though host is very sophisticated, all it does is maintain its internal equilibrium state; facilitate consciousness with all information and tools so that consciousness can make decisions and execute actions.

REASONING NORMALIZATION

If mankind deals with its surroundings as static and permanent, what difference does it make? How does the new knowledge about the continuum and the body help their daily activities?

What is the main purpose of the normalization? Or why should anybody follow the normalization procedures?

These are the questions we will look into, to justify the need for the normalization, before going into the normalization methodologies.

JUST COMMON SENSE

Normalization is neither spiritual nor divine but it is based on just common sense. That is it defines the behaviour of the consciousness, the observer (and the actor), based on the true nature of the field, the continuum, and the nature of the host — the human body.

Normalization is based on simple common sense — If everything we observe is virtual and exists only for a fraction of time then how our behaviour should be towards them is explained by these methodologies.

WHAT YOU SEE IS WHAT YOU GET

All the stress, unhappiness and hostility that we observe in human society is due to the misunderstanding of the surroundings, the field and the host. There is no surprise why humans behave in the way they do, because humanity thinks the universe as something permanent and takes everything as personal.

This can be explained very clearly using the dog–mirror example (*fig*. 5.15).

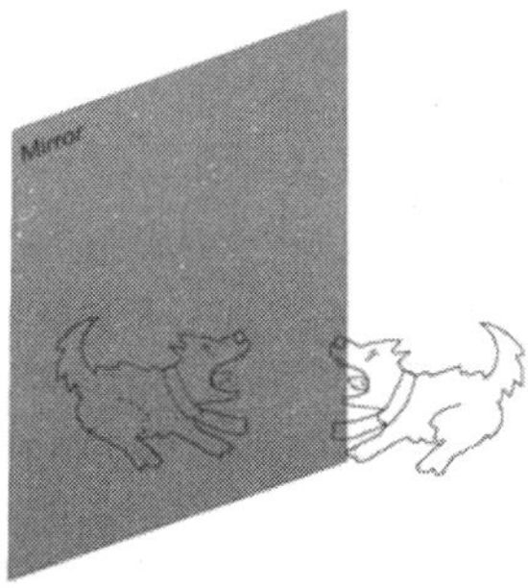

fig. 5.15: Dog understanding

As the dog sees itself in the mirror, its reaction will be similar to the way it behaves towards some other dog. Since dog cannot understand the nature of the mirror and the reflection, it reacts in a non-intelligent way.

Similarly, by assuming the projection of the fraction–living continuum onto the brain, a person without proper knowledge about the continuum will react in a non-intelligent way and bring stress to the body.

Bottom line, there is nothing mysterious why humanity is suffering. It is just because they are not able to completely understand the true nature of the entities and their interactions around them.

HAPPINESS IS ONLY A SIDE EFFECT

But again the main purpose of normalization is not to make human societies happy and stress-free.

The main purpose of normalization is to make the humanity to realize the true nature of their surroundings. By targeting the source of all the problems — the ignorance of humanity — normalization procedures provide a new way of living which is simple, normal and logical. Rest is just a side-effect of this change in attitude towards worldly entities and their interactions.

MAINTAINING HOMOEOSTASIS CONDITION

Like any other (non-living) physical system, the human body always tries to stay in the equilibrium state. A small deviation from this state means the internal systems of the human body shall secrete harmones to bring the system back into a balanced state.

As consciousness perceives the surroundings as static and permanent field, it gets involved with the virtual activities. By taking the virtual entities personally, it imposes unintended stress on the body. The human body fights back to maintain the homoeostasis condition bringing the stress and fatigue on to itself.

The main purpose of consciousness normalization is to

educate the consciousness so that it leaves its host in the equilibrium state for the most part by treating the worldly entities and activities as they are, virtual and fraction existing.

FROM VIRTUALITY TO REALITY

The following steps propose few methodologies for the consciousness to achieve the goal of staying in reality.

These methodologies are designed addressing consciousness, where the word "consciousness" can be read as an individual or a person. The methodologies are discussed in detail in the normalization book (Jata 2011).

FRACTIONAL CONTINUUM

The most important concept consciousness has to remember is that the continuum even though looks so vast, exists only for a fraction of time.

> The main result of this feature is — consciousness can be less critical of objects and activities as it knows they after all exist for only a fraction of time and it doesn't make any sense to lose mind on them.

ONE ENTITY

The consciousness must remember that even though the objects in the three-dimensional world look separated and distinct, they all are one single entity.

> The main result of this feature is — as the continuum brings variety in the entities and activities, the consciousness will not get perturbed, or deals with them less critically and relatively equally by treating them as one.

NO TIME-LINE

The human brain stacks up the information from multiple spans of continuum to give an illusion of continuous, permanent entities and activities.

Continuous time-line results continuous actions. Continuity in actions creates the illusion of larger span of continuum. By denying the continuity, the consciousness can see the objects and activities being "floating" through the structure of the continuum at each fraction of the continuum time-line.

DETACHMENT IS THE KEY

Fig. 5.16 shows the span of the continuum as perceived by the consciousness due to getting involved with the virtual world.

So how to minimize the *māyā* (trickery of the universe) and start living in reality?

First of all, it needs to be understood that it is impossible to live in present, as the present is so small. It is also impossible not to think about past and future as the basic human activities spread across trillions of continuum spans.

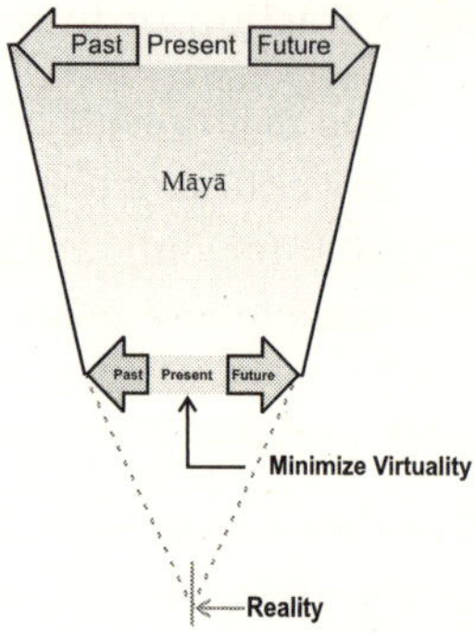

fig. 5.16: Towards reality

The detachment that is discussed here is not by rejecting the activities, but by realizing the true nature of them, thereby changing the approach towards them.

Since it is impossible to refrain from doing things, the actions must be performed by always keeping in mind and remembering, the true nature of the entities and activities. It is not about not doing actions but it is about doing them or dealing with them by realizing the true nature of these entities and activities.

Fig. 5.17 illustrates the process of detachment. As consciousness detaches from the activities and the entities, it performs the day-to-day actions without being critical of the result from them but will perform them on as needed basis.

But the nature of the detachment is not based on rejecting anything consciousness dealing with but realizing the true nature of the surroundings.

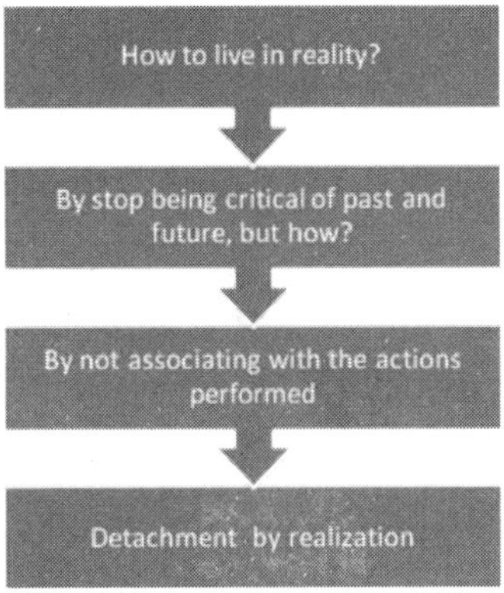

fig. 5.17: Detachment process

- Since the external environment and the activities are virtual, there is nothing to take seriously.
- Since continuum exists for only a fraction of time, the activities and the consequences become trivial.
- Since we cannot avoid doing activities, do them without getting attached them — like watching a movie.

By applying the methodologies explained so far, the state of the consciousness can be illustrated. Since the methodologies addressed the external environment, consciousness realization is illustrated using the wake state:

- All the entities will become one, the single entity — the continuum.
- Consciousness treats them as they should be treated, a single entity. *Fig*. 5.18 shows the individual entities in the same colour as the continuum to indicate that consciousness stops seeing separate and distinct three-

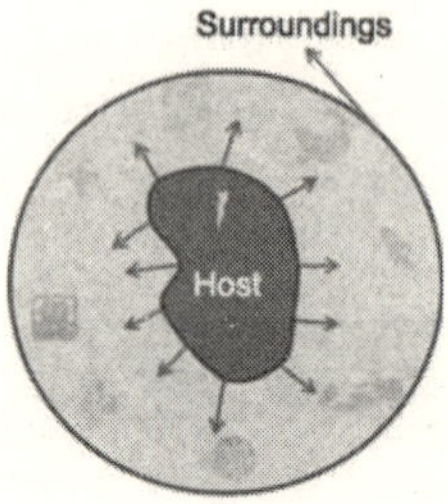

fig. 5.18: Waking state: one entity

dimensional world and realizes the single entity — the four-dimensional continuum.

- But consciousness cannot ignore them completely.
- The surroundings have to be dealt with as needed, to protect the host and to perform activities.
- Even though detached from the entities, activities must be performed in a detached manner or in a manner we deal with virtual entities.
- Ignore but be aware as needed just like watching the virtual movie.

Fig. 5.19 shows the interactions between consciousness and surroundings with dashed lines. The dashed lines indicate the waking state of the consciousness being both the performer and the observer.

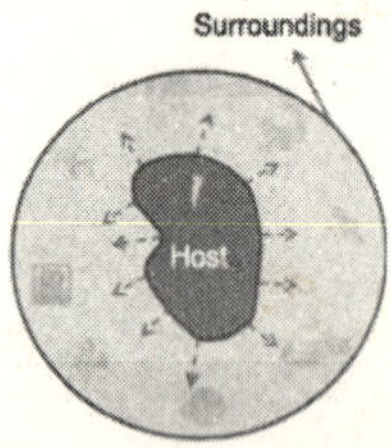

fig. 5.19: Detached

Again remember, all the methodologies are just based on common sense. Consciousness normalization is just a change in the attitude based on the nature of the continuum. And the detachment from the activities does not imply not doing anything, but it indicates an ideal behaviour towards virtual activities and their virtual consequences.

VIRTUAL HOST

Just like the entities in the continuum are virtual, the host of the consciousness is also a virtual entity.

The following methodologies help consciousness to realize the virtual host and stay in reality.

> By extending the same principle to the host, i.e. by realizing the virtual nature of the host and that consciousness is not host, consciousness will interact with the host on as needed basis too. Which means consciousness will not misuse the facilities provided by the host, but will use them only as needed.

INDEPENDENT HOST

The host does not require consciousness intervention for the most part unless there is somebody need or there are some changes from the external surroundings.

> Consciousness must realize that it is not completely defined by the host and host is sophisticated enough to operate by itself for most part.`

PRESERVE HOMOEOSTASIS HOST

Few examples, what helps maintain the homoeostasis nature of the host is: eat healthy, exercise, sleep as needed, good body hygiene, etc.

By realizing that consciousness is not the host, consciousness must recognize the importance of the host, maintain it as carefully as possible and execute the actions related to the host only on as and when needed basis.

> As the consciousness detaches from the virtual continuum, it is a must for the consciousness to remember that the host itself is virtual and it is sophisticated enough to maintain itself for the most part.

By applying the methodologies discussed, the detached state of consciousness from the host is illustrated using the dream state. Here the dream state is picked as it is this state during which consciousness identifies with the host.

- The host and the rest will become one, the single entity — the continuum.
- Consciousness treats the host as it should be treated, a virtual and part of the single entity, the continuum. *fig.* 5.20 shows the host in the same shade as the continuum to indicate that consciousness stops seeing separate and distinct three-dimensional host and realize the single entity — the four-dimensional continuum.

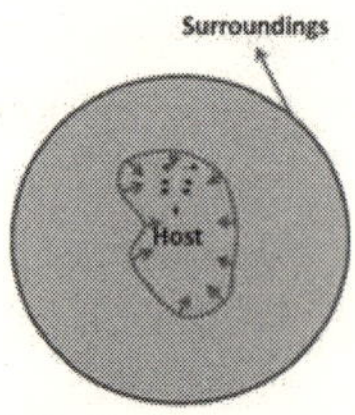

fig. 5.20: Consciousness is not host

- But consciousness cannot ignore the host completely.
- The host has to be dealt with as needed, to protect the host and to perform activities.
- Even though detached from the host, activities must be performed to maintain the host and perform the activities.
- Ignore but be aware as needed.

Fig. 5.21 shows the interactions between consciousness and its host with dashed lines. The dashed lines indicate the dream state of the consciousness being both the performer and the observer or a complete detached dream state.

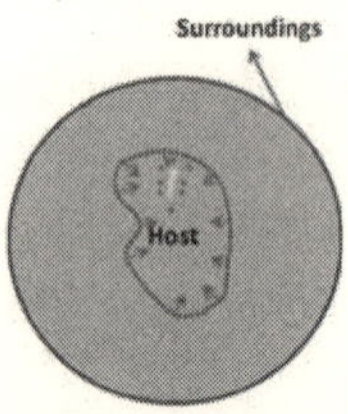

fig. 5.21: Detached dream state

NORMAL STATE

Combining the detached waking state and the detached dream

state will give the resultant balanced state for the consciousness.

Detached State

Fig. 5.22 shows the detached state of the consciousness while performing the daily activities.

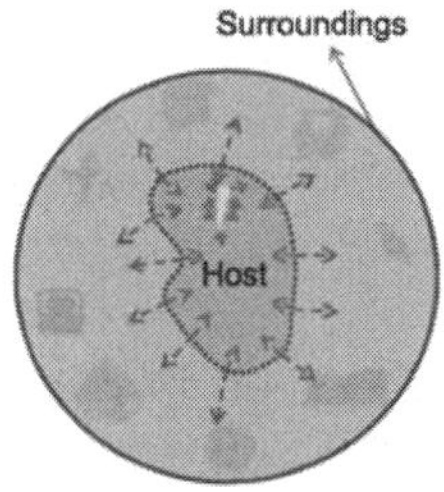

fig. 5.22: Detached state

Consciousness and Continuum

Consciousness realizes that even though it is associated with the host, or it cannot survive without it, the host is part of the continuum. Since the continuum is one single entity, host is also continuum; consciousness sees all the entities, including the host as one single entity.

Fig. 5.23 illustrates the singleness of the continuum and consciousness as being embedded in the continuum.

This concept is not that radical. As everything on the movie screen is part of one entity, all the entities in the continuum are one. Since consciousness is embedded inside the continuum, it is actually not limited to the host but the whole continuum is its host.

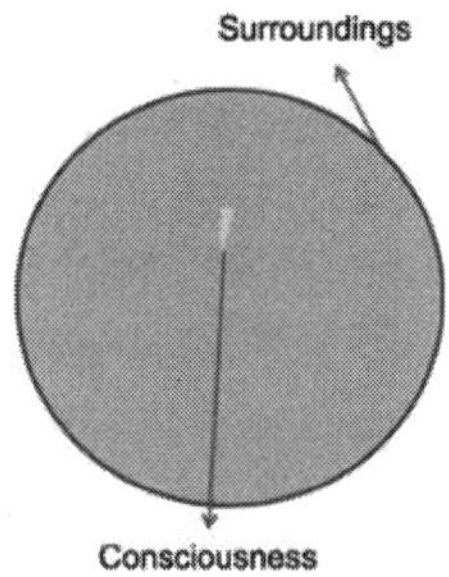

fig. 5.23: Consciousness and continuum

Consciousness Normalization

Finally summarizing the behaviour of consciousness with the four-dimensional continuum:

- The detached state is called as the most "normal state" of the consciousness.

- This state is achieved by treating the entities as they should be, i.e. the continuum is not permanent but just fractional existence and entities are just part of one single entity.
- This state is achieved by treating the host as it should be, i.e. the consciousness is not host and host is part of continuum.
- In this state, the consciousness is neither fond of nor ignorant of its surroundings.
- All the actions will be performed without any attachments to them, including body activities as the consciousness detaches from them by realization.
- Erases the boundaries between the host and surroundings by realizing that everything is nothing but one entity, the continuum.
- This state is the most "normal state" as in this state consciousness does not get involved with the virtual entities.
- Since consciousness does not get personal, it will not pose unnecessary stress on the host, thereby leaving the host in homoeostasis condition.

CONCLUSION

The methods discussed in this section help consciousness to stay away from the virtual nature of the continuum and realize the true nature.

In fact, all the pieces needed for achieving the normalization are already there. All the non-living entities already are in sync with the continuum nature. Even the human body is already tuned for normalization and it facilitates powerful mechanisms which are either completely automated or semi-automated.

All that consciousness needs is to just behave in accordance with the field, the continuum and the host.

The methodologies explained are not divine, spiritual or superficial, but they are just based on common sense. They are all derived from the nature of fraction-existing, four-dimensional continuum. These methodologies only educate consciousness about the entities it is dealing with.

Along with these methodologies, by following the best practices, as suggested by normalization guide (in the normalization book (Jata 2011)), one can easily come out of the old habits and achieve his/her natural and normal behaviour.

Measuring Normalization

This section will discuss about measuring or observing the consciousness normalization quantitatively.

FLIGHT OR FIGHT

Flight or fight is the first impulsive feeling expressed by consciousness while facing a situation. Consciousness makes the decision whether to fight the situation or take to flight to avoid the situation. An abnormal consciousness, i.e. the consciousness that is not normalized, considers the three-dimensional world as static and permanent. Due to this it takes every situation more seriously and gets involved with the virtual entities and builds up attachments. Due to these attachments emotions, expectations and tensions in the body build up.

Due to the lack of understanding of the nature of the continuum, an ignorant consciousness will tend to be over-critical of everything, extend control beyond the host, inflict stress on the body and feel the pain of exhaustion.

Once the host is toppled by the alarm signal from the stress mechanism, the homoeostasis mechanism will kick in to balance the physiological changes in the body. This continuous balancing act leads to exhaustion of the host.

BALANCED LOOP

As the sophisticated human body collects all the sensory information, it is the consciousness that makes the decision using the central nervous system. An individual without the proper knowledge about the continuum may deal with the information in an abnormal way ending up with attachments and emotions. These attachments and emotions fire up the alert signal in the central nervous system (CNS).

Fig. 5.24 illustrates the balancing loop of the human host to maintain the homoeostasis state. As the CNS fires up the neuron activity due to unwarranted emotions and attachments, the homoeostasis mechanism (HSM) will activate the autonomic nervous system (ANS) to balance the system. During this process, neuro-transmitters like adrenaline and other stress hormones are released. These fight or flight hormones are responsible for various physiological changes in the body like — heart rate increase, blood pressure, shaking, etc.

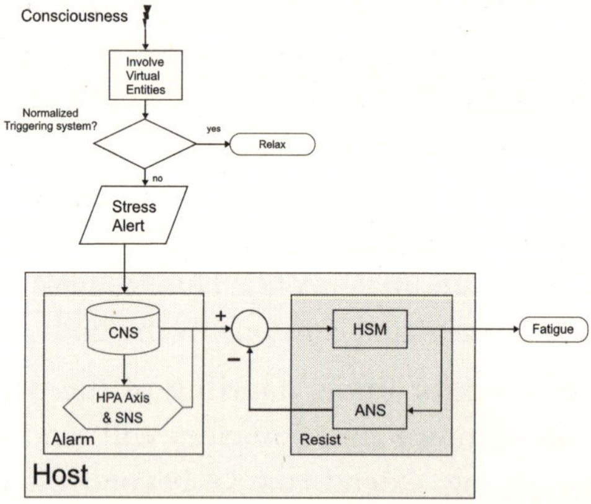

fig. 5.24: HSM balancing

Since the host cannot control the initiation or interactions with the surroundings by the consciousness, it simply reacts to what is commanded. In *fig.* 5.24, once the alarm block sends the signal, the resistance will be activated to minimize the impact, thereby preserving the safety of the host.

MEASURING TECHNIQUES

Various methods are available to visualize the structure of the brain or monitor the activity of the CNS. The physiological changes in the body can also be monitored using these methods.

- Magnetic resonance imaging
- Functional magnetic resonance imaging
- Positron emission tomography
- Electroencephalography

What to Measure

Various tests were conducted on the three states of consciousness using the above-mentioned techniques. Studies are also being conducted on meditation to understand how meditation affects the human body and the CNS.

Table 5.4 shows the measurements and observations for the three states. Roughly 75 per cent of the sleep is non-REM sleep and the remaining is the REM sleep.

Fig. 5.25 shows the physiological changes during REM and non-REM sleep. As shown in the figure, the heart rate, eye movement and the respiration rate are lower during the deep sleep mode indicating calmness.

Fig. 5.26 shows the EEG rhythms during the three states. In recent times, stage 3 and stage 4 are combined together. Delta rhythms correspond to brain activity during the deep sleep state of consciousness.

Measuring Normalization

The normalization procedures suggested help consciousness to behave as it should, thereby leaving the host in a relaxed state. Moreover, the normalization procedures are scientifically derived and they can be implemented at all the times with reason, and without having to sit in one place or, without concentrating on one entity.

Table 5.4: Three States of Consciousness

Behaviour	*Awake*	*Non-REM*	*REM Sleep*
EEG	Low voltage, fast	High voltage, slow	Low voltage, fast
Sensation	Vivid, externally generated	Dull or absent	Vivid, internally generated
Thought	Logical progressive	Logical repetitive	Vivid, illogical, bizarre
Movement	Continuous, voluntary	Occasional involuntary	Muscle paralysis; movement commanded by the brain but not carried out
Rapid eye movement	Often	Rare	Often

Source: adopted from Table 19.1 (Bear et al. 2007).

A completely normalized consciousness can achieve a totally relaxed host without any stress and anxiety. The normalized state can indirectly be observed by monitoring the physiological changes in the host.

As shown in *fig*. 5.25, in the deep sleep state, the heart rate, the body temperature and blood pressure are lower when compared to the corresponding readings in the other two states of the consciousness. The other indication of subdued brain activity is the EEG rhythms of the brain in the range of delta wavelength illustrated in *fig*. 5.26.

Since in normalized state the host will be in a relaxed state, similar physiological readings (as in non-REM state) can confirm the achievement of consciousness normalization.

Conclusion

NORMALIZED STATE CONCLUSIONS

Here are the characteristics of the fourth state derived from the consciousness normalization procedures:

- Fourth state is the normal state because in this state the consciousness does not associate with illusion or virtuality.

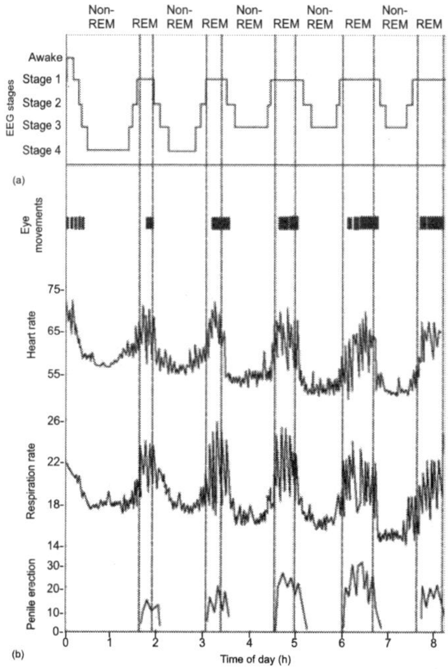

fig. 5.25: Body changes

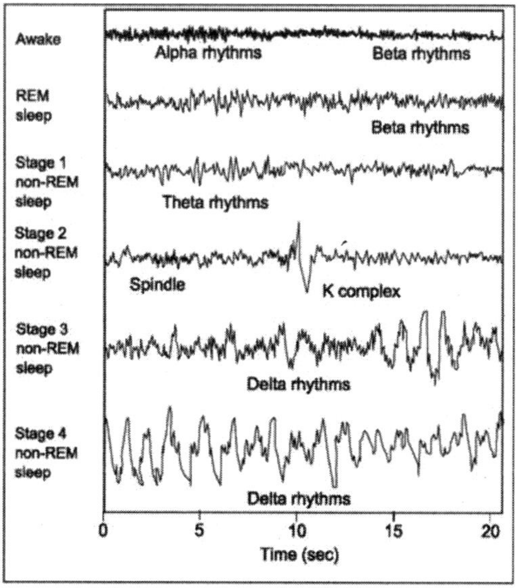

fig. 5.26: EEG rhythms

- By realizing the oneness of the continuum, in this state consciousness does not limit itself to the host, but separates itself from the host and looks at everything as one.
- In this state the body will be completely relaxed as there won't be any unnecessary intervention from the consciousness.
- Normalization is not for the host, but it is for the consciousness to leave the host alone.
- There is no mystery in the normalization procedures; no spirituality, no divinity, no religion or no godly concepts. These methodologies are designed based on pure logic and common sense.

MASTER OF VIRTUAL ENTITIES

The entities we observe are virtual, exist for only a fraction of time and they have no significance. Consciousness must realize this all the time and be the master of the virtual world. What we see — the entities and their interactions — have no value to them as they last for only a fraction of time. What matters for the consciousness is to rise above the virtuality and lead a normalized life.

CYCLE OF CONSCIOUSNESS

The future of evolution of humanity is nothing but the normalization of their behaviour towards the virtual continuum. In other words, rationalizing the human behaviour based on the fact that what is real and what is virtual.

The reunion of consciousness with itself or with the continuum (as it is part of continuum) can be summarized as follows:

- As the consciousness entered the virtual continuum, started dealing with the three-dimensional objects and activities, it evolved the host into the sophisticated human body. As human CNS helped consciousness to deal with the virtual world and the virtual time-line, consciousness lost itself in the trickery of the virtual field.

- As consciousness realizes the true nature of the continuum, starts dissociating itself from the virtual world with the help of normalization procedures, it gains back its oneness, and achieves the destination — going back into the reality where it came from, the real and never changing continuum; and still be able to deal with the virtual world effectively.

SIMPLE PHYSICS

Normalization methodologies are derived based on simple laws of nature, or on the basic laws of physics.

The normalization procedures simply help consciousness in understanding the nature of continuum and lead a logical and relaxed life based on the basic laws of nature. The stone, the normalization state is achieved by simply complying with the nature of the field, the nature of the host and letting them do their job.

There is nothing new for the consciousness to do, all it needs to do is leave the system alone without unnecessary interference and in return enjoy the ride of life.

SCIENCE AND SPIRITUALITY

Normalization procedures are the future of the humanity. Normalized way of living is the only way humanity can escape the stressful life and lead a logical, relaxed, happy life.

This is the first time in the history of humanity a way of living is designed based on simple and scientific fundamentals.

With the concepts discussed in this report, there will not be any conflicts between science and spirituality. As explained by normalization methodologies, they are completely connected as science leads consciousness to spiritual freedom.

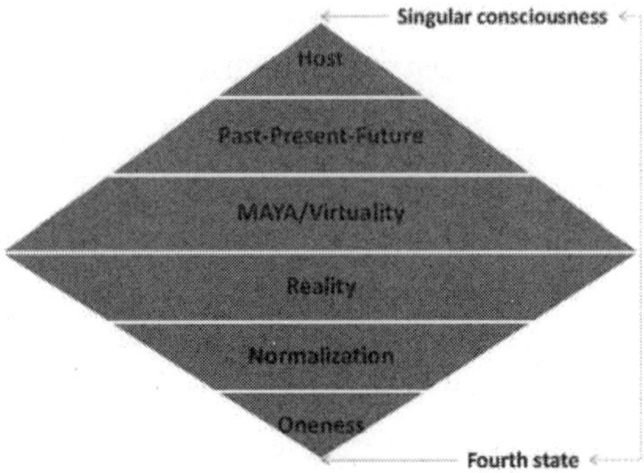

fig. 5.27: Consciousness normalization

FINAL WORDS ON NORMALIZATION

Setting aside all the discussion, ask yourself some basic questions:

- What is there to be particular about when everything exists only for a fraction of time, like a single frame in the movie?
- When everything is one single entity, like a computer screen, where is the identity?
- Consciousness normalization is the most important task to achieve for any individual because this is the only exit from the illogical, stressful lifestyle.
- Without the normalization, consciousness continuously struggles with the virtual entities and exhausts the host.
- Normalization procedures are easy to remember, are logical, scientific and are the future of the humanity as long as the continuum exists.

AFTER NORMALIZATION

The question to ask is, once a complete normalization is observed by an individual for a long time, what kind of benefits or changes can be observed? Is it possible to extend the lifespan, as the host is always maintained properly without any stress and with good care?

Just like the way humans evolved from their predecessors, is it possible for the host to genetically imprint normalization onto genes? As the consciousness gets normalized, is it possible that some change in physiology of the body can occur?

Normalization process by itself is justifiable as it makes the consciousness behaviour normal, as it should be behaving, towards the virtual continuum and the virtual host.

SANĀTANA DHARMA IS THE FUTURE OF HUMANITY

As explained in the report, the concepts of Sanātana Dharma can be summarized as follows:

- The four-dimensional continuum is identified as *Paramātmā*.
- Tesseract representation is provided.
- *Paramātmā* features are explained in multiple ways and compared with lightning to highlight the fractional existence.
- Since everything is fractional, the concept of *māyā* is introduced.
- Since the host is also virtual, the concept of *ātmā* (consciousness) is introduced.
- Since everything is virtual, the concept of detachment is introduced.
- With detachment and looking everything as one single entity, the concept of merging with the *Paramātmā* is introduced.
- Since normalization is the only way, the realization of *Paramātmā* or the four-dimensional and singular nature of continuum is considered as the utmost responsibility of humanity.

- *Turīya* state is introduced as it is the next evolutionary state of consciousness.

Sanātana Dharma is the only future of humanity because logically it is the only way of living that keeps the humanity from virtual way of living.

Note: This is a report on the books *Consciousness Normalization* by Ravi Jata ISBN 978-0-9839644-0-7 (Jata 2011) and *Brahmajignasa* by Ravi Jata ISBN: 978-0-9839644-1-4 (Jata 2012). Most of the text and figures are referred from these books. This report is an abridged version and does not contain all the details mentioned in these two books.

References

Bear, Mark F. Connors, W. Barry and, Michael A. Paradiso, 2007, *Neuroscience: Exploring the Brain*, s.l. : Lippincott Williams & Wilkins.

Chanana, Raj D., 1983, *Rig-Bhashya-Sangraha*, 3rd edn., New Delhi: Munshiram Manoharlal, pp. 250-60.

Coxeter, H.S.M., 1973, *Regular Polytopes*, 3rd edn., New York: Dover Publications.

Gardner, Martin, 2001, *The Colossal Book of Mathematics*, New York: W.W. Norton and Co., pp. 162-74.

Jata, Ravi, K., 2012, *Brahmajignasa*, Saint Paul: Stratum Publishing Company.

———, 2011, *Consciousness Normalization*, Saint Paul: Stratum Publishing Company.

Prabhupada, Bhaktivedanta Swami, 1997, *Bhagavad Gita As It Is*, Mumbai: Bhaktivedanta Book Trust.

Sarvananda, Swami, 1985, *Isavasya Upanishad*, Chennai: Sriramakrishna Math Printing Press.

———, 1986a, *Kaṭha Upaniṣad*, Chennai: Sriramakrishna Math Printing Press.

———, 1985b, *Muṇḍakopaniṣad*, Chennai: Sriramakrishna Math Printing Press.

———, 1985c, *Kena Upaniṣad*, Chennai: Sriramakrishna Math Printing Press.

———, 1986b, *Māṇḍūkya Upaniṣad*, Chennai: Sriramakrishna Math Printing Press.

Tyagisananda, Swami, 1986, *Śvetāśvatara Upaniṣad*, Chennai: Sriramakrishna Math.

6

A Proposed Model to Quantitatively Assess Consciousness

Employing Time and Temperature Relationship to Mind

Bal Ram Singh, *Rishi Bal Singh*,
and *Sukalyan Sengupta*

Abstract

A scientific analysis of consciousness and the mind remains elusive due to the lack of a measurable quantity that can adequately reflect these domains. It is, therefore, important to develop some *ab initio* efforts that allow the measurement of tangible parameters explaining the mind and consciousness. We hypothesize that consciousness is a general concept of existence, is universal, and includes any and all things, both known and unknown. Further, awareness of this consciousness begins by first limiting it to one or more of its infinite dimensions, including space and time, which then enables the mind to actually perceive it.

We have devised a model which creates a theoretical basis for using time and temperature to examine consciousness through mind's perception. The model projects vibrations to be the characteristic feature of consciousness. Assuming consciousness to be represented by space vs. time, a signal in space will increase in frequency with a decrease in time, and as the time reaches zero or infinity, the signal will become all-pervasive, and will represent reaching of a state of complete consciousness. Contraction and expansion of time, is thus, a defining factor to attain consciousness. We have used the kinetic theory of ideal gases to define time in terms of temperature, and propose to define a mind or consciousness coefficient which may be resolved at an individual level to define one's consciousness.

Introduction

A FUNDAMENTAL question of the sense of self has remained in the human mind throughout history and it has been addressed variously by different cultures and traditions over a long period of time. While it is generally considered to result from self-reflection through introspection, in actuality it is a continuous phenomenon, perhaps since the beginning of life and primordial mind. In a recent model by Swanson (2003) for the functional organization of the brain, there are three types of inputs (*fig.* 6.1) to motor neuronal systems which determine expressive behaviour: sensory, intrinsic, and cognitive. The sensory inputs result in reflexive responses, cognitive inputs produce voluntary responses, and intrinsic inputs act as control signals to regulate the behavioural responses.

Interestingly, the origin of the intrinsic inputs is not understood, and is recognized primarily from the observation that during embryogenesis and foetal development, the motor neurons are developed before sensory neurons and start operations without any sensory signals. This led to the conclusion that there must be a mechanism, internal to motor neurons, which begins operating before any sensory development. In other words, external signals to neurons are not necessary for them to have a sense. This observation leads to one of the fundamental questions of consciousness, at least at the biological level: Are the cells conscious of themselves? This question can be explored further by asking a more

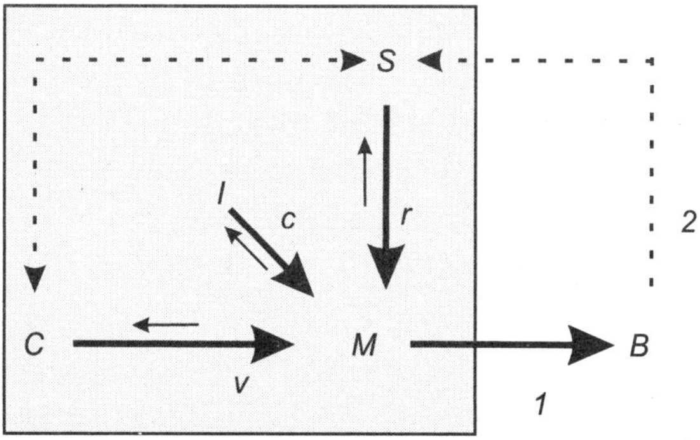

fig. 6.1: A model of the functional nervous system's basic diagram. The model information flow (yellow box) assumed that behaviour (*B*) is determined by the motor system (*M*), which is influenced by three classes of neural input — sensory (*S*), intrinsic (*I*), and cognitive (*C*). Sensory inputs lead directly to reflex responses (*r*), cognitive inputs mediate voluntary responses (*v*), and intrinsic inputs act as control signals (*c*) to regulate behavioural state. Motor system inputs (1) produce behaviours whose consequences are monitored by sensory feedback (2). Sensory feedback may be used by the cognitive system for perception and by the intrinsic system to generate feedback (e.g., positive and negative reinforcement/pleasure and pain). The cognitive, sensory, and intrinsic systems are all interconnected, hence arrowheads at the ends of each dashed line within the nervous system box (adapted from Squire et al. 2008)

fundamental question — are molecules or atoms conscious of themselves? The answer, of course, will depend on the basic definition of consciousness itself.

Consciousness is generally defined as wakefulness, alertness, perception of surroundings and response, etc. Before elaborating on these ideas, it is important to consider this process fundamentally while accepting these general definitions. At the human level, and also at the animal level, the wakefulness, alertness, etc. make perfect sense and are intuitively acceptable for defining consciousness. However, when considering plants, bacteria, and virus, one has to resort to the basic biological processes at the micro scale, such as signal transduction, membrane transport, where one can find ample examples of reception of signals and responses, which could run in a time scale of seconds to milliseconds. Viral consciousness, in that sense, can be considered at nanosecond levels, which would be the time scale involved in their binding to target cells and tissues; processes involved at molecular and atomic levels go further down to picosecond levels or even lower.

Consciousness, considered from atomic to human levels, thus can be given a general term of reception and response process of matter, which is related to scale of time involved in the processes that reflect "consciousness". In contrast to sub-nanosecond phenomena, there are mega and cosmic levels of consciousness which involve thousands to billions of years and include the origin of universe, origin of life and evolution of life, etc. The time scale used here is important to emphasize the relationship between these processes, human perceptions, and memory, all of which are objects of consciousness.

An interesting experiment conducted by Paul Ekman[1] elucidates the importance and linkage of time, consciousness, and perception. His experiment focuses on the most neurologically evolved sensory system: the visual system. It utilizes a videotape in which a series of faces show a variety of expressions very briefly. The challenge is to identify whether you've just seen, the facial signs, for instance, of contempt or anger or fear. Each expression stays on the screen for just one-fifth of a second in one version, and for one-thirtieth of a second in another — so fast that you would miss it if you blinked. Each time the participant must select which of the seven given emotions he or she has just seen. The ability to recognize fleeting expressions shows an unusual capacity for accurate empathy. Such expressions of emotion — called micro-expressions — happen outside the awareness of both the

---

1 See http://mindful.org/the-science/neuroscience/a-monk-in-an-mri

person who displays them and the person observing. Because they occur unwittingly, these ultra-rapid displays of emotion are completely uncensored, and so reveal — if only for a short moment — how the person truly feels.

Ekman found that a Tibetan monk and another advanced Western meditator were two standard deviations above the norm in recognizing these super-quick facial signals of emotion, albeit the two subjects differed in the emotions they were best at perceiving. They both scored far higher than any of the five thousand other people tested. "They do better than policemen, lawyers, psychiatrists, customs officials, judges — even Secret Service agents", the group that had previously distinguished itself as most accurate.

Thus, understanding the perception of time itself is therefore, a valid subject of inquiry to bring consciousness into the quantitative domain of science and technology.

Perception of the World Beyond "I"

To perceive, understand, and comprehend processes and phenomena beyond open perceptual range (sub-seconds for processes and years for memory), a human being uses cognitive and imaginative processes which are still subject to intrinsic and sensory processes of the brain, as recognized by modern science, and mind — which still remains elusive to most mainstream scientific fields. However, there are perceptions beyond the physical realm which still need to be considered. The following three levels of human state of consciousness provide the basic framework of examination in this chapter to develop a conceptual basis of further elaboration of time in relation to mind.

LEVEL 1: "I" AND THE WORLD ARE SEPARATE

(1-a) Information processing and awareness can be explained through neurobiology. According to Crick & Koch (1990), Crick

(1994), Koch (2004) and others, certain 35-75 Hz neural oscillations in the cerebral cortex seem to be correlated with awareness in a number of different modalities and a mechanism of binding (synchronization of separately represented pieces of information) has been hypothesized. In our model, the "consciousness coefficient" can be used to "control" the time needed to resonate cortical oscillations with a chosen object of awareness. Thus, the self chooses what objects enter into the "frame of awareness" and how long they stay; therefore, a framework independent of physical time can be developed. How we develop this control is discussed later. However, it is logical that "multitasking" or "continuous partial attention" are antithetical to this process, and concentration or focus will be needed to manipulate neurobiology. Davidson's work with Tibetan Buddhists (Lutz et al. 2004) has demonstrated the ability of Lamas with many years of meditative practices (> 10,000 hours) to generate gamma waves ($\approx$ 40 Hz) when asked to meditate on positive emotions.

In another study, Newberg et al. (2001) conducted similar research on the subject of meditation and the brain activity of Tibetan monks. According to the subsequent findings, gleaned from scans of each monk's brain during meditation, an increase in activity was found around the frontal region of the brain, in which attention on specific tasks are processed; on the other hand, a decrease in activity was found around the area at the back of the brain, where one's processing of orientation and spatial awareness occur.

Regarding the results of his study, Newberg found that:

> During meditation, people have a loss of the sense of self and frequently experience a sense of no space and time and that was exactly what we saw.

He concluded,

> When someone has a mystical experience, they perceive that sense of reality to be far-greater and far-clearer than our everyday sense of reality. Since the sense of spiritual reality is more powerful and clear, perhaps that sense of reality is more accurate than our scientific everyday sense of reality.

This sense of reality is best seen in the first meeting in Dakshineshwar that Swami Vivekananda (then Narendra Nath Dutta) had with Sri Ramakrishna in November 1881. When Narendra asked Sri Ramakrishna, "Sir, have you seen God?", SRK immediately and unequivocally replied, "Yes, I have seen God. I see Him as I see you here, only *more clearly*".

(1-b) The use of psychedelic drugs can cause time dilation and "expansion of consciousness" (Bentov 1979) but we do not agree with "expansion of awareness" because this state is purely reflexive in nature and the agency of self-control is absent.

(1-c) State of Dream — In REM sleep (the state where dreams occur) time dilation is reported but once again this state is outside of the purview of self-control.

Thus, in level 1, manipulative time dilation is only possible if we are in the waking state and can increase the gap between two thoughts/sensory perception modalities. This is possible through *yoga*/meditation (*yogaś cittavṛtti-nirodhaḥ*).

LEVEL 2: SUṢUPTI

In *suṣupti*, deep sleep or non-REM sleep, Ādi Śaṅkarācārya states [*Vivekacūḍāmaṇi* (171); Turiyananda 2003],

> *suṣupti-kāle manasi pralīne*
> *n 'aiv 'āsti kiñcit sakala-prasiddheḥ*

> At the time of *suṣupti*, when the mind is absorbed in *māyā* — when there is no mind — there is nothing. I and the world are not distinct, but this is still a level of ignorance (*avidyā*), in that the experiencing self is not aware that it is but a part of the Whole.

LEVEL 3: TURĪYA

The fourth level of awareness (beyond waking, dream, and deep sleep), *turīya*, can be characterized as:

- Purest state of awareness
- Consciousness transcends limitations of space, time and causation
- Consciousness merges with the Ultimate
- Not a state of knowledge but pure experience

Thus, any model to quantify consciousness, must be able to encompass the above states of awareness and, more importantly, also be able to control the agency of space and time without any external input.

Concept of Space: Origin of Space

Space is a product of the Big Bang Singularity, but it can be posited that the Big Bang Singularity itself is a state before which everything was pure existence and pure consciousness (*ānīda-vātam*), without vibration, as described in the Hymn of Creation, the Nāsadiya-Sūktam (*Ṛgveda* X.129).

- There was neither non-existent nor existent . . .
- That one thing, breathless, breathed by its own nature.

It then existed without vibration. *Prāṇa* (primal energy) existed then but there was no motion in it. Then when the *kalpa* begins after an immense interval, the *ānīda-vātam* commences to vibrate, and blow after blow is given by *prāṇa* to *ākāśa*. The atoms become condensed, and as they are condensed different elements are formed.

All that is power in the universe, manifesting as force or attraction — or even thought — is *prāṇa*.

prāṇa + *ākāśa* = Universe

Every manifestation of power is *prāṇa*

Every material manifestation is *ākāśa*.

This vibration then becomes the first source of time and space, as we understand the gap between two events or gap between two points, respectively. Quantitative measure of these terms runs from zero to infinity, a concept well examined in the Vedas. An article by D.R. Sharma (2003) explains that although mathematics regards zero and infinity as the two extremes of the number system, they are very different from ordinary numbers. By their very definition numbers are quantitative. They have meaning only in relation to some objects. By themselves they mean nothing; five has nothing five-like, it is only a word. But zero and infinity have no quantitative aspect to them, they can be referred to only qualitatively. Each is an indivisible whole. They do not depend on other objects for their meaning. In other words, they are absolute, not relative. Any number multiplied by zero or infinity loses its identity, it merges into the multiplier. Any number divided by zero yields infinity, but divided by infinity yields zero. Zero and infinity exist side by side. The same idea is expressed in Vedānta from a different perspective.

Sharma continues as follows:

> In Vedānta zero and infinity are regarded unknowable and therefore explaining or defining them is out of question. Even in mathematics zero and infinity cannot be arrived at; a variable can only tend to be either. Looking for the smallest number greater than zero or the largest number smaller than infinity is futile. For any unknowable we cannot say what it is; it can be described [is] only in terms of what it is not. That is why in answer to the question "what is the Ultimate Reality (*Brahman*)" the Vedas say "not this, not this (*neti neti*)".

Sharma points out that the Sanskrit word for zero and all its equivalents in other languages is *śūnya*.

> It is far more comprehensive in meaning than zero; it means zero only in the context of numbers. Its general meaning is void or emptiness, a concept that the mind cannot grasp unless it is itself empty (the initial premise of Yoga philosophy). The same is true for infinity. In Sanskrit there is no single word for infinity. Several words like *anādi* (without beginning), *ananta* (without end), etc. are used.

According to Sharma, "Upaniṣads characterize *Brahman* as smaller than the smallest and larger than the largest" (*Śvetāśvatara Upaniṣad* 3.20; *Kaṭhopaniṣad* 1.2.20). In the usual sense, in which the words small and large are used, this statement may seem to be a contradiction in terms. This is because our concept of small and large is tied to the perception of space and time. Small and large are a pair of opposites, which, according to Vedānta are two faces of the same coin. Small is contained in the large and large is contained in the small in a latent form. As Kabīr, one of the great Indian mystics said:

> The drop merges in the ocean, where do you search for the drop? The ocean merges in the drop, where do you search for the ocean? The end of the search is the realization that each is in the other (Sharma 2001). A singularity brings zero and infinity together.

The cosmological singularity combines zero space–time and infinite energy (Sharma 2003).

> In Vedānta the primordial state of existence of the Reality is an infinite emptiness (*śūnyatā*). When the mind is completely empty and one transcends space–time in deep meditation, one is able to realize this state. However, this emptiness is not nothingness or non-existence. It is pervaded by energy and consciousness. One arrives at the same concept coming from the quantum physics; the quantum void is not empty. There are myriads of virtual particles appearing and disappearing constantly and it is filled with an all-pervading cosmic energy. This state is the substratum of all

> existence and is called the "ground" by mystics as well as physicists (Krishnamurti and Bohm 1985). The dictionary meaning of singularity is "the state of being unique or only one of its kind". In that sense we can regard the Ultimate Reality as the cosmic singularity. In quantum physics one cannot see or probe "naked singularity"; it is enveloped by the event horizon. From Vedānta's point of view we cannot "see" the Reality because it is veiled by *māyā*. In cosmology space-time goes to zero at the singularity but the concept is still wrapped up in space–time; it is defined as a point singularity. In case of the Ultimate Reality the singularity is devoid of any geometrical connotation. It is beyond the grasp of the rational mind.

Existence of anything is only represented in space–time. Anything that has name and form is therefore limited in space–time. How one can transcend space–time? It is physically impossible, but physical body is not the true identity of oneself. There are subtle bodies — ethereal, astral, mental, and spiritual which can transcend the space–time limitation to which physical body is bound. According to Sharma (2003):

> space–time implies separation — distance between two points or two events. When every object is viewed as an event and the object vanishes, space–time also vanishes. The vanishing of the subject–object duality essentially means that the person has merged with the universe. In this state one has transcended space–time and zero and infinity have come together. The Upaniṣads describe this as the state in which a person sees everything in himself and himself in everything. — *Īśopaniṣad* 5.6

Concept of zero and infinity is inseparable, and is consistent with the Vedāntic idea of undivided wholeness which expresses the omnipresence of the Supreme or *Brahman*. *Īśopaniṣad* says:

> That (*Brahman*) moves and remains still at the same time;

That is near as well as far; That is inside as well as outside.
— *Īśopaniṣad* 5.6

Everything in this universe and beyond is in fact the manifestation of the *Brahman*, thus nothing is beyond it. One of the great sayings (*mahāvākyas*) of Vedānta is *aham brahmāsmi* (I am Brahma; *Bṛhadāraṇyaka Upaniṣad* 1.4.10 of the *Yajurveda*). "I" refers here not to the body or name but the all-pervading power of Brahmā. If "I" is everywhere at the same time, then for "I" there is no space and there is no time. Still there is movement because Brahmā is consciousness (*prajñānam brahma* — "Consciousness is *Brahman*" (*Aitareya Upaniṣad* 3.3 of the *Ṛgveda*), which is dynamic (Sharma 2003). However, now the concept of movement is totally different; space and time are not involved in it. The concept of non-locality in modern quantum physics comes close to this viewpoint.

According to Vedānta, Brahmā is pure consciousness, which is in every living and non-living being. So everything that we perceive or know is through that Brahmā; That is the ultimate knower. The subject–object duality precludes the possibility of being conscious of our own consciousness in the same way as we are conscious of an object (Sharma 2003). How can the knower be known — is a famous exclamation in Vedānta (*Bṛhadāraṇyaka Upaniṣad* 2.4.14).

Two questions thus arise in deriving quantitative measures of observations made:

1. Can objective and subjective observations be examined scientifically? This is an important question, as according to the Vedic literature the whole cannot be known under the dual nature of the observer and the observant.

2. Can there be a quantitative measure of the consciousness itself? This is an important issue if we assume answer to the question 1 above is a "yes". Since

it is the attainment of infinity with the consciousness that brings about zeroness to the vibrations, there is a possibility that intermediate stages between the beginning and the end of this attainment can be estimated.

Modern Scientific Approach and Vedic Ideas

Modern science, as we know today, has had many of its roots in the Vedic and Upaniṣadic knowledge, although its current concepts were deliberately dissociated even from philosophical bases, mostly due to historical reasons. After European Renaissance during 1400s to early 1500, there was a major introspection in the Western world, largely under the shadows of the Church, which controlled the Western society in those times (Kirkpatrick 2002). India — where the Vedic practices, even if these were not perfect or genuine, still dominated the society — was under the foreign rule, with very limited freedom to deal with such issues. European Renaissance — looking back to the Greek philosophies and even Roman Empire — provided inspiration. Consequently, scientific theories and interpretations had a clear bias with respect to philosophy and culture.

A caution is in order to apply modern scientific methods and fields of learning to understand ancient knowledge, especially when there may be major cultural and philosophical differences between the society where these scientific tools are developed and the society of the ancient knowledge. Therefore, it is important to examine the basic premises in the Western philosophy or science, and compare them with some of the fundamental assumptions in the Indian philosophical systems.

Western thoughts (Agrawal 2001) hold that (1) man cannot know metaphysical truths by direct experience, (2) even if they can be known, so far no man has known them, and (3) being pure speculations, various schools of Indian thoughts, like the speculative system of the West, must be mutually contradictory, and that if one is true then all the others must be false. Presuppositions for Indian thoughts are: (1) Man can know metaphysical truths directly, (2) there have been such men and women and there may still be seers with such knowledge, (3) seers or *ṛṣis* teach metaphysical truths after knowing them directly, and (4) while all the *ṛṣis* know the same truths, they teach these truths in different standards represented by the texts of the corresponding schools (Agrawal 2001).

The definition of modern science is a systematic analysis of an issue, problem, or the object, which provides consistent results. Moreover, scientific truth is something that cannot be falsified. There is no need for direct evidence to prove it true. Both these elements of modern science are fortified by the requirement of objectivity in pursuit of scientific enquiry and experiments.

On the surface it appears there could be a major contradiction between the Vedic knowledge and the Westernized modern science. However, in practice, at least at the fundamental level, science follows the Vedic ways of pursuing the truth. For example, scientific pursuit embraces the idea of knowing the truth in more ways than one, and actually scientists find more than one way to prove their points or ideas, be it more than one technique, organisms, or even researchers. Therefore, in the current time of scientific predominance of learning, it is incumbent upon scholars to use the science itself to examine some of the intricacies of the ancient knowledge.

Some of the great luminaries of the twentieth-century science have clearly stated the Vedic thoughts to exactly explain the highest ideas of the modern science. Erwin Schrödinger, one of the major pillars of modern quantum mechanics writes (Schrödinger 1964):

> Within a cultural milieu (*kulturkries*) where certain perceptions (which once had or still have a wider meaning among other peoples) have been limited and specialized, it is daring to give to (my conclusion arrived at by putting together quantum mechanics with biology) the simple wording that it requires. In Christian terminology to say: "Hence I am God Almighty" sounds both blasphemous and lunatic. But please disregard these connotations for the moment and consider whether the above inference is not the closest a biologist can get to, proving God and immortality at one stroke.
>
> In itself, the insight is not new. The earliest records, to my knowledge, date back some 2,500 years or more. From the early great Upaniṣads the recognition *ĀTMAN = BRAHMAN* (the personal self equals the omnipresent, all comprehending eternal self) was in Indian thought considered, far from being blasphemous, to represent quintessence of deepest insight into the happenings of world. The striving of all the scholars of Vedānta was, after having learnt to pronounce with their lips really to assimilate in their minds this grandest of all thoughts.
>
> Again, the mystics of many centuries, independently, yet in perfect harmony with each other (somewhat like particles in an ideal gas) have described, each of them, the unique experience of his or her life in terms that can be condensed in the phrase: *DESU FACTUS SUM* (I have become God). . . .

In Einstein's view, experiencing the notions of space and time lie at the foundations of his theory of relativity (Ananthu 1992). According to Einstein, "For a physicist, the distinction between past, present, and future is an illusion, though a stubborn one". And this was not anything less than a true science (Ananthu 1992). Einstein had a very strong opinion on the subject of mystic, science and religiousness (Ananthu 1992):

> The most beautiful and [the] most profound emotion we can experience is the sensation of the mystical. It is the sower

> of all true science. He to whom this emotion is stranger, who can no longer wonder and stand rapt in awe, is as good as dead. To know what is impenetrable to us really exists, manifesting itself as the highest wisdom and the most radiant beauty which our dull faculties can comprehend only in their primitive forms — this knowledge, this feeling is at the centre of true religiousness.

At the same time, there have been prominent scientists who completely believe in the reductionist theory of view to understand and explain the nature of living and non-living, and show even the modest contempt towards religious way of realizing the truth. Nobel laureate Francis Crick who, on one hand, was so much affected by Erwin Schrödinger's famous book, *What is Life*, that he changed his field from physics to biology (Satya and Umesh 1992), nevertheless took a very different point of view of understanding life, primarily by reducing everything to the smallest molecules. Crick (1994) observes:

> Most of the religious beliefs we have today originated in a time when the earth, while a small place by our standards, was then thought of as being very large, even though its exact extent was unknown. Any one person had direct knowledge of only a tiny part of it. It was not implausible to believe that this large earth was the centre of the universe and that man occupied the leading place in it. The earth's origins seemed lost in the mists of time and yet the span of time thought to be involved, while it seemed long in terms of human experience, we now know to be ridiculously short. It was not implausible to believe the earth was less than 10,000 years old. We now know its true age is about 4.6 billion years. The stars seemed far away, fixed perhaps in the spherical firmament, but that the universe extended as far as it does — more than 10 billion light years — was almost inconceivable. (An exception has to be made here for certain Eastern religions, such as Hinduism, that take

pleasure in inflating times and distances for the sheer joy of it.)

It is obvious that there are scientists like Crick who are very hostile to religious groups of all sorts without willing to examine the ancient knowledge itself. However, even he had to acknowledge the contribution of the Vedic knowledge of the concept of time, even if reluctantly.

With the realization of the limitation of human sensual faculties, scientists have adopted models to examine and understand the matter as a substance. These models have been developed, and are acceptable (Barton 1997) because they (a) describe the new in terms of the old, (b) express unfamiliar in terms of the familiar, (c) relate the abstract to the concrete, or (d) integrate several previous models in an intuitively satisfying way.

It is notable that with the application of these models we acknowledge as scientists that we are not even dealing with reality, just a simple perception of it. At the best, these objects in the solid world around us are really only images or approximations around us (Stapp 1982). Our scientific theories are just theories, their relationship to reality again depends on the perception. Even the instrumental measurements are based on a perceived reality. Echoing these sentiments, Stephen Hawking said in a 1994 lecture: "I don't demand that a theory correspond to reality because I don't know what it is" (Hawking and Penrose 1996).

Given the above limitations, it is still possible to model reality in terms of the modern scientific languages. Chemistry certainly represents a language of science that can be used to model more than the two other fields of natural sciences, biology and physics, primarily due to their complexity and abstractness, respectively.

Hypothesis

Consciousness is a general concept of existence of any and all the things known and unknown, and is universal. The consciousness in its ideal and complete form is experienced equally by everything, including human beings. Awareness of the consciousness occurs by limiting the consciousness to one or more of the infinite dimensions of the consciousness, including space and time.

Any one of these limitations is reflected in mind which acts as the relativist of a human being. Mind is a form of energy that changes frequently with time, space, and any other changes in the universe by being an integral part of the consciousness. The relationship of mind with time and space opens the possibility of its examination by the tools of modern science, which is primarily the study of matter as a function of time and space.

In order to examine the consciousness, one has to create a model for consciousness which could be based on intuitive assumptions yet lead to measurable parameters to validate the model and create an understanding of the consciousness in a scientific manner. We make a model of the consciousness by assuming the following:

1. Consciousness is all-pervasive.
2. Expression of consciousness can be represented by vibrations.
3. Different things may have different frequencies of vibration.
4. Vibrational frequency can be defined as periodic change in signal in any of the infinite dimensions of consciousness with respect to another dimension of the consciousness (e.g., space vs. time, energy vs. time, space vs. energy, etc.).

5. Vibrations are natural to all matter and non-matter components of the universe and para-universe.
   - Assuming the consciousness to be represented by space vs. time, a signal in space will increase in frequency with decrease in time (*fig*. 6.2), and as the time reaches zero, the signal will become all-pervasive, and will represent reaching of the consciousness. Similarly, with increase in time (reaching infinity) wavelength of the signal will increase to become all-pervasive (*fig*. 6.2, right panel), again reaching the consciousness. In either case, the amplitude of the signal remains constant irrespective of variation in time. Contraction and expansion of time is thus a defining factor to attain consciousness.

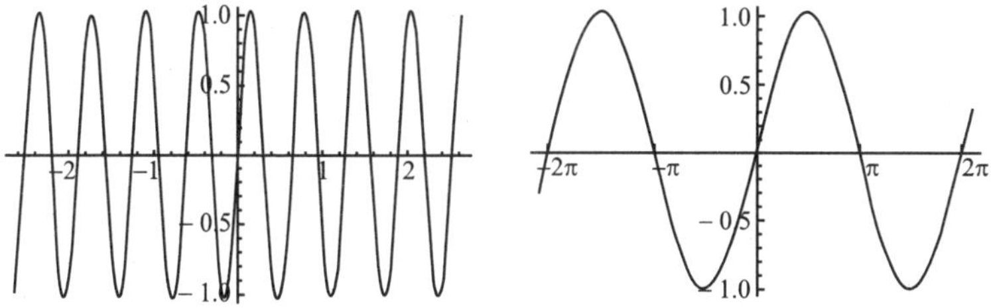

fig. 6.2: Signal changes with high (left panel) and low (right panel) frequencies, maintaining constant amplitude irrespective of the frequency

Time in a Circle as a Representation of Consciousness and Mind

The correct geometric model for time has been argued over for many years by various philosophers and thinkers, most seeing time as either linear or circular. While the theory presented here does not by any means end this debate, it does present a new look at time in terms of how it is perceived, and presents new ideas as to why time may travel in circular, rather than linear, motion.

The main purpose of this model of time is to show how the mind's perception of time changes with respect to one's consciousness. Its basis lies in the fact that a certain amount of physical time may be perceived by us to be much longer or shorter than it actually is. For example, Einstein once conducted an experiment dealing with this very issue. For the first part of his experiment, he sat on a hot plate for what he thought was an hour; the actual amount of time he spent sitting on the hot plate was closer to a minute and a half. Afterwards, he sat with a pretty girl for what he felt was a minute, in actuality sitting with her for fifty-six minutes. Thus, our mind's perception of time can often be vastly different than the physical measurement of time.

Einstein's experiment and our observations kindled in us an inspiration to create a model that could map this phenomenon. The circle, which represents time as a whole, is the basis for the model. The angle measure of any piece taken of the circle represents a physical amount of time; so, for example, if the entire circle were to represent one hour, the 90° portion (shown in *fig*. 6.3) would be the equivalent of fifteen physical minutes.

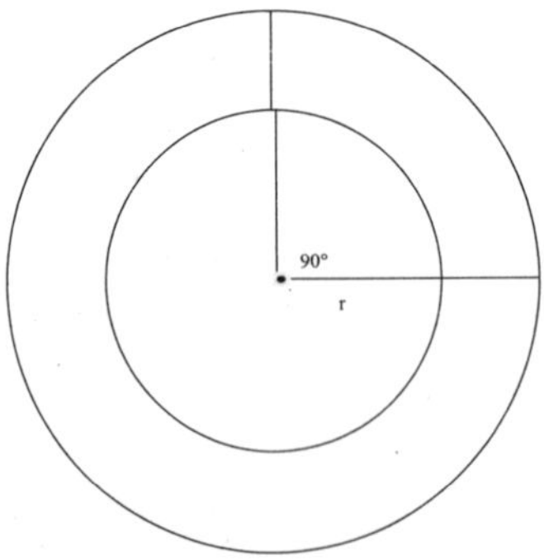

fig. 6.3: The circle of time showing an inner circle and an outer circle. According to our model, the 90° angle shows two perimeter arcs representing two different levels of consciousness indicated by the respective radii.

The radius of the circle that one uses depends upon the consciousness of the mind. A mind completely focused or engrossed in an activity will be represented by a circle closer to the centre, while one less focused, distracted, or unengaged will be represented by a much larger circle, being placed farther from the centre.

The mind's perception of time corresponds to the length of the arc created on that mind's specific circle. The less engrossed one's mind is, the longer its perception of time becomes. When one's mind is less engrossed, it is represented by a circle farther away from the centre, and while the angle and actual measure of time stay the same, the mind's perception of time, and the length of the arc, both increase.

This model created to map our perception of time also fits the Hindu view of time and consciousness. First of all, Hindu philosophy sees time as circular and cyclical, such as the cycle of *yuga*s and the cycle of birth and rebirth. Thus the circular model presented fits well.

Further, in Hindu philosophy it is believed that the further one gets from this physical world, and the closer one gets to total consciousness or God, the less time and space matter or even exist. This also suggests that as time and space matter less, one is closer to total consciousness. This also means that the centre of the circle is in fact total consciousness, or Brahmā.

Using the centre of a circle to represent Brahmā or total consciousness further fits Hindu philosophy, which says that while Brahmā is infinite and everywhere, He is also simultaneously intangible, unable to be seen, and thus to the human eye existing as a result of faith in His existence. The centre of a circle is in similar situation: it cannot actually be found, but it is known to exist. The moment any physical point is picked as the centre of a circle, that point takes up physical space and thus it itself becomes another circle, whose centre must be found. In this way the centre, being infinite yet not able to be found, easily fits with the Hindu description of total consciousness.

In addition, interestingly, Hinduism sees physical, or what we might call "real" time to actually not be real at all. In Hindu philosophy, all worldly dimensions, including that of time, are part of *māyā*. In this way, a mind less bound by the restrictions of worldly time is also closer to the Hindu definition of reality, and thus at a higher level of consciousness.

Altogether, using the relationships this model creates between consciousness and the mind's perception of time, it is possible to use the model in order to take a measurement of one's level of consciousness and its relationship specifically to time.

Temperature and Time, Mind and Einstein's Theory of Relativity

The easiest way to make a connection of time and temperature with mind is the observation Einstein made in 1938 (Einstein 1938):

> When a man sits with a pretty girl for an hour, it seems like a minute. But let him sit on a hot stove for a minute and it's longer than any hour. That is relativity.

The important thing to note here is the interconnectedness of temperature with the observation of time. This is equivalent to the level of mental tolerance of adverse conditions by the people, something that can be empirically measured. This also allows the individualization of the observation which is subjective. It would, however, be measurable and comprehensively by others.

Measurement of Consciousness: Time in terms of Temperature

Once a connection is established that it is the mind that is

fig. 6.4: A cartoon showing experimental finding of Albert Einstein on time dilation and mind (taken from S. Mirsky, 2002, "Einstein's Hot Time", *Scientific American* (September), p. 102)

responsible for the time-temperature perception, one can derive the time in terms of temperature, allowing a direct measurement of mind (which is the limited version of consciousness) by making assumptions about the instrument of mind, which is brain, and the lifetime experience of human being.

- Kinetic theory of gases:
- $Vrms = (3RT/M)^{1/2}$;

  $Vrms$ – root mean square velocity

  » R – Universal gas constant

  » T – Temperature, K

  » M – Molar mass

- Defining $Vrms$ as d/t (d, distance in metres; t, time in seconds), and re-arranging the equation, one can get

  – $t = (d^2M/3RT)^{1/2}$

  - assuming d and M to represent constants, the expression can result in:
  - $t = C/T^{1/2}$, where C is **Consciousness Coefficient** that can be estimated for each individual or entity experimentally.
  - $C = (d^2M/3R)^{1/2}$.
- For humans, M would be 1.4 kg, average mass of a human brain and d would be 7.0956×10^{17} metres (average lifespan, 75 years, in seconds x speed of light, 3×10^8 m/sec). $R = 8.3145$ J/K.mol.
- An unevolved human, defined totally on material basis, would have a Consciousness Coefficient, C, of 1.68×10^{17} m.K.mol/J.kg, where m is metre, K is Kelvin, mol is moles, J is Joules, and kg is kilogram.

The C could become a powerful tool to examine individuals for the evolution of their mind, and it could perhaps be useful to their medical and social care. This being so practical, more data can be collected for evaluating the scientific nature of this subjective science, and compare it with the objective science being practised in the modern world.

Conclusion

The series of assumptions and logical and mathematical derivations suggest that it is possible to accept a subjective way of learning and examine it through modern scientific tools. The Vedic ideas of zero and infinity provide a holistic and dynamic range for quantification of consciousness. A vibrational model of the beginning of universe, representing both time and space, can allow the formulation of mathematical equations representing knowledge and ignorance to explain human diversity. A circle of time model (*fig*. 6.3), provides a testable hypothesis that the time span, in fact, represents the distance from the total reality, an idea that resonates with the *yuga* concept. Using heat perception on one hand, and an engrossed attachment on the another, one can evaluate the involvement of mind in realization of time. At a gross level, the brain and lifespan of a human being allow one to calculate Consciousness Constant which can be experimentally tested and verified to evaluate the proposed model.

References

Agrawal, M.M., 2001, *Six Systems of Indian Philosophy*, Delhi: Chaukhamba Sanskrit Pratishthan.

Ananthu, T.S., 1992, "Does God Play Dice?", in *Philosophy of Science*, ed. Narlikar et al., Delhi: Munshiram Manoharlal, pp. 39-48.

Barton, A., 1997, Chapter 1, *States of Mind States of Matter*, Bristal: Institute of Physics Publishing.

Bentov, I., 1979, *Stalking the Wild Pendulum*, NY: Bantam.

Crick, F. and C. Koch, 1990, "Toward a Neurobiological Theory of Consciousness", *Seminars in the Neurosciences*, **2:** 263-75.

Crick, F., 1994, *The Astonishing Hypothesis: The Scientific Search for the Soul*, New York: Scribners.

Einstein, A., 1938, "On the Effects of External Sensory Input on Time Dilation", *J. Exothermic Science and Technology*, **1**(9).

Hawking, S.W. and R. Penrose, 1996, "The Nature of Space and Time", *Scientific American*, **275:** 44-49.

Kirkpatrick, R., 2002, *European Renaissance 1400-1600*, New York: Longman.

Koch, C., 2004, *The Quest for Consciousness: A Neurobiological Approach*, Englewood, CO: Roberts and Company.

Krishnamurti, J. and D. Bohm, 1985, *The Ending of Time*, New York: Harper and Row.

Lutz, A., L. Greischar, N.B. Rawlings, M. Ricard and R.J. Davidson, 2004, "Long-term Meditators Self-induce High-amplitude Synchrony during Mental Practice", *Proceedings of the National Academy of Sciences*, **101:** 16369-373.

Newberg, A., E. D'Aquili and V. Rause, 2001, *Why God Won't Go Away: Brain Science and the Biology of Belief*, Ballantine.

Rai, D.R., 2003, "Cosmic Singularity: A Vedanta Perspective", *Metanexus News*, 6 March.

Satya, Y.S. and G. Umesh, 1992, "Chance and Determinism in Physics", in

Philosophy of Science, ed. Narlikar et al., Delhi: Munshiram Manoharlal, pp. 96-117.

Schrödinger, E., 1964, *What Is Life?*, Cambridge: Cambridge University Press.

Sharma, D.R., 2003, "Zero Infinity And God", Sulekha.com (http://dharambir-rai-sharma.sulekha.com/blog/post/2003/03/zero-infinity-and-god.htm), 2 March.

Sharma, R.K. (ed.), 2001, *Kabir Granthavali* (*Complete Works of Kabir*), Allahabad: Oxford University Press, pp. 169-70.

Squire, L.R., D. Berg, F. Bloom and S. du Lac, 2008, *Fundamental Neuroscience*, 3rd edn., Academic Press.

Stapp, H.P., 1982, "Mind Matter and Quantum Mechanics", *Foundations of Physics* **2:** 363-99.

Swanson, L.W., 2003, *Brain Architecture*, Oxford: Oxford University Press.

Turiyananda, Swami, 2003, *Vivekacūḍāmaṇi of Śrī Śaṅkarācārya* (Sanskrit Text with Transliteration, Translation and Index), Ramakrishna Math.

7

Vedic Perspectives on Acoustics

M.G. Prasad

Abstract

Sound plays a major role in various aspects of life. The field of acoustics deals with generation, transmission and reception of sound. Sound in modern scientific terms refers to the pressure waves in air received by human beings in the audible range of generally 20 cycles/sec to 20,000 cycles/sec. However the infra- and ultra-sound refer to the frequencies below and above the audible range respectively. Acoustics has also received a very high importance in Vedas and the Vedic literature. The classical literature of Hinduism or Sanātana Dharma is rooted in the Vedas. It is well known that the Vedas are collection of *mantra*s. In the Vedas and the Vedic literature several terms such as *nāda* and *śabda* are used. These terms cannot be translated directly as sound, as they refer to vibrations including both physical and spiritual aspects. However the *nāda* and *śabda* include the human auditory range sound. In particular human speech, vocal and instrumental music as they lie within the audio range have been discussed significantly in Vedas and Vedic literature. This paper presents the Vedic perspectives on acoustics. The Vedic fourfold model for generation of speech is discussed with applications of general sound. The acoustical aspects of speech, music, language, literature and spirituality are discussed. The paper also includes discussions of the Vedic view in relation to views of modern acoustics.

Introduction

VEDA refers to the body of ancient knowledge that deals with physical, psychological and spiritual aspects of life and the cosmos. The word *veda* etymologically relates to the verbal root *vid* meaning "to know". However, it is interesting to note that the Vedas also refer to *mantras*, which means "chants that convey knowledge". The Vedic chants have acoustical characteristics such as phonetics, tonal quality, length measure, effort, etc. These Vedic chants have been orally transmitted with precision from master to disciples over thousands of years. It is well known that the precision in the oral transmission of chants, music and speech cannot be surpassed through writing. Thus acoustics of human speech is very important. Also it is well known that the faculty of speech is revered as Goddess of Speech (Vāgdevī). The *mantra* from the *Yajurveda* says:

> *devīṁ vācamajanayanta devāḥ, tāṁ viśvarūpāḥ paśavo vadanti* ।
> *sā no mandreṣamūrjaṁ duhānā dhenurvāgasmānupasuṣṭutaitu* ॥

The meaning of this *mantra* could be summarized as:

> Creator Lord Brahmā and other *devatās* brought forth Vāgdevī (Goddess of Speech) to manifest communication. All beings use their abilities to communicate through Vāgdevī. May Vāgdevī, like an all-desire fulfilling cow, bless us with food, strength and faculty of speech!

It is quite difficult to precisely define and describe the origin of language and human speech. However, it is known that Vedic chants and music in India have existed from very ancient times. Sir C.V. Raman, in his article titled "The Acoustical Knowledge of Ancient Hindus" (Raman 1922), says that:

> It would form a fascinating chapter of history to try and

trace the gradual development of musical instruments and musical knowledge, from the rhythmic chanting of the *Ṛgveda* in the ancient home of the Aryan race to the Indian music of the present day.

Importance of Acoustics

In modern times the importance of acoustics is very well known. The field of acoustics deals with generation, propagation and reception of sound. The importance of the field of acoustics can be seen as shown in the *fig*. 7.1, which shows the role of acoustics in arts, life sciences, earth sciences and engineering.

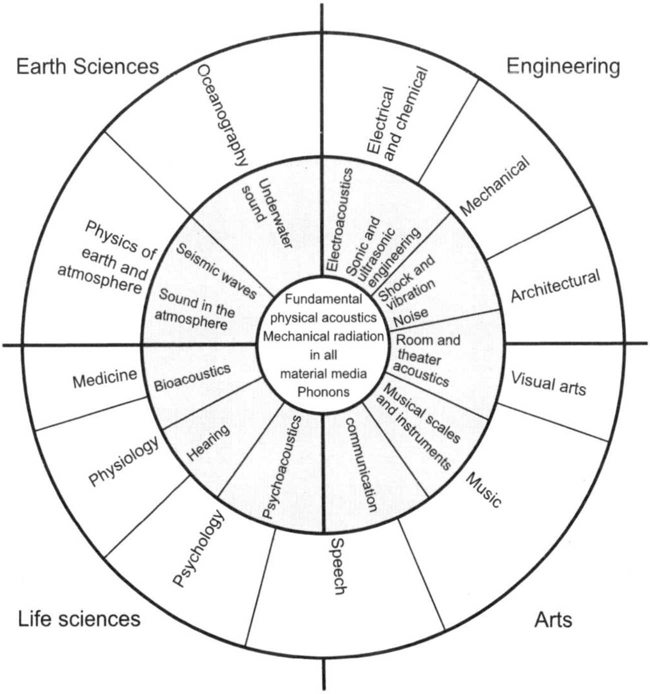

fig. 7.1: Lindsay's wheel of acoustics showing the presence of acoustics in various fields

It is interesting to note that acoustics has high importance in the Vedas and Vedic literature, which are the classical source literature for Sanātana Dharma, also popularly known as Hinduism. The Vedas and Vedic literature, which is voluminous as shown in *fig.* 7.2, deals with all aspects of life including sciences and arts.

For example, in reference to *fig.* 7.2, *Gandharva Veda* deals with music and fine arts, and Chandas deal with science of prosody and its knowledge helps in melodious pronunciation

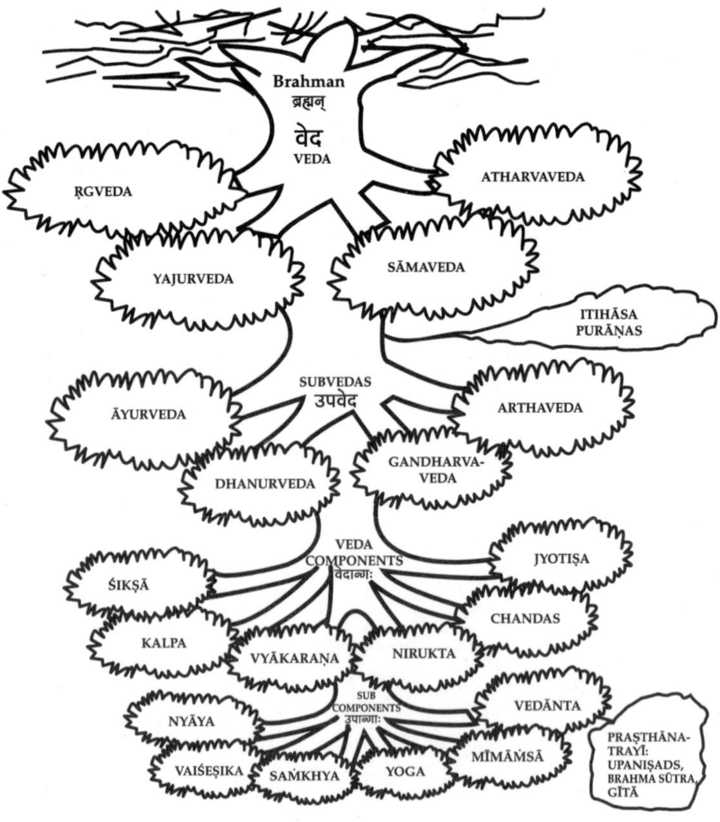

fig. 7.2: Integrated Vedic knowledge and science represented as an inverted tree

of *mantras*. It is well known that the four main Vedas (*Ṛg*, *Yajur*, *Sāma* and *Atharva*) are collections of *mantra*s, which have been orally transmitted through the ages. The proper recitation of *mantra*s requires proper knowledge of acoustics for their recitation. It is interesting to note that UNESCO has also recognized the tradition of the oral transmission of Veda *mantra*s as below:

> the verses of the Vedas were traditionally chanted during sacred rituals and recited daily in Vedic communities. The value of this tradition lies not only in the rich content of its oral literature but also in the ingenious techniques employed by the brāhmaṇa priests in preserving the texts intact over thousands of years. To ensure that the sound of each word remains unaltered, practitioners are taught from childhood complex recitation techniques that are based on tonal accents, a unique manner of pronouncing each letter and specific speech combinations.
>
> — UNESCO Proclamation 2003
> (Intangible Cultural Heritage — ICH)

The precision in oral transmission from teacher to student is directly related to acoustics. The six acoustical factors involved in precise Vedic chanting as per *Taittirīya Upaniṣad* are, pronunciation of letters (*varṇaḥ*), their pitch (*svaraḥ*), timing or duration (*mātrā*), force (*balam*), melodious articulation (*sāma*) and combined sound effects (*santānaḥ*).

Although sound is generated by varieties of sources in nature, the sound produced by human beings is of chief importance because of its role in communication. In the Vedas and Vedic literature, the terms *nāda* and *śabda* are referred to indicate both audible and inaudible sounds. This can be referred as *Sound Field*. The production of sound from human beings as speech and music are part of the *nāda* and *śabda*. However it is important to remember that the Vedic terms *nāda* and *śabda* cannot be translated as sound in the audible range (of 20 to 20000 Hz). The various manifestations of sound in the world that are generated and propagated become part of the sound field. The sound field as *nāda* and *śabda* has received utmost importance in Vedic literature. Śārṅgadeva, the author of *Saṅgīta Ratnākara*, a treatise on classical Indian music, says:

nādena vyajyate varṇaḥ padaṁ varṇāt padāt vacaḥ ।
vacaso vyvahāro 'yaṁ nādādhīnamato jagat ॥

> The alphabets have manifested from the *nāda* (sound field). The words are made up from alphabets and from the words speech comes out. Worldly transactions are carried out through the speech, and thus, the world is dependent on *nāda*.

In a similar way, in the words of a *yogī* (seer) Śrīraṅga Sadguru (1983-2008):

> Just as from seed comes naturally sprout, branch, leaf, flower, unripe fruit and full fruit, likewise, starting from spiritual light as seed, *nāda*, *svara* and *akṣara*s have developed into the various aspects of knowledge.

The sound field as *śabda* and *nāda* takes us to the origins of the universe. In the *Amr̥tabindu Upaniṣad* (Aiyar 1997), one of the minor Upaniṣads, the following verse says:

dve brahmaṇi veditavye śabda brahmaparaṁ ca yat ।
śabda brahmaṇi niśṇātaḥ param brahmādhi gacchati ॥

> Two *Brahmans* to realize, namely: *Śabda* and *Param Brahman*. One who has realized and is well versed in *Śabda-Brahman* will realize *Param Brahman*.

Thus, it is seen that the knowledge of sound field as *śabda* and *nāda* is a prerequisite to the spiritual fulfilment of realization of *Param Brahman*. Also the sound field as the revered *Nāda-Brahman* in addition to its being all-pervasive in the world is also the source of Bliss. Śārṅgadeva gives it utmost importance

in the Vedic literature. Śārṅgadeva, the author of *Saṅgīta Ratnākara* (Shringy and Sharma 1978) says:

> *caitanyaṁ sarvabhūtānāṁ vivṛtaṁ jagadātmanā* ।
> *nāda brahma tadānandaṁ advitīyam upāsmahe* ॥
>
> We worship the *Nāda-Brahman* second to none, which is blissful and is in all beings as consciousness has manifested itself as the universe.

The scriptures also refer to two types of *nāda*, namely *anāhata* and *āhata*. The *anāhata* refers to self-existing unproduced sound field without any vibratory cause and is experienced only in spiritually focused deepest states of Yoga. The *āhata* refers to the sound field produced with vibratory cause. The discussion in this paper deals with produced sound field, namely *āhata nāda*.

Although sound (*āhata nāda*) is generated by both human beings as well as non-human sources in nature, the ones produced by and the ones that affect human beings manifest as speech, music, literature and spirituality (Yoga), as shown in *fig*. 7.3.

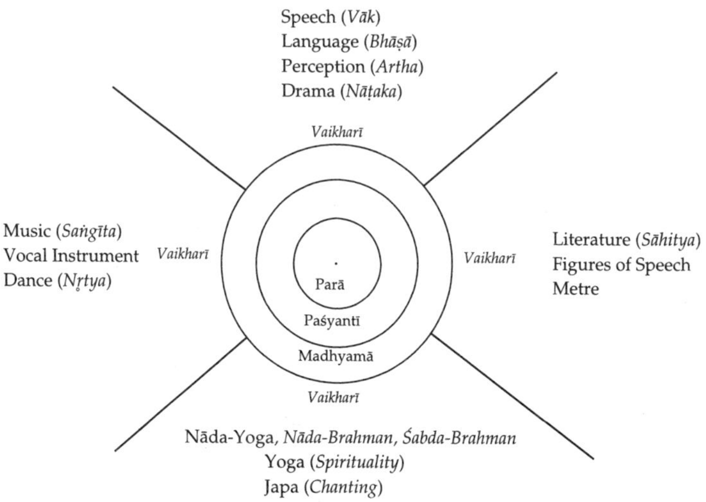

fig. 7.3: Representation of the four fields of sound at Vaikharī level

In the Vedic literature, major emphasis is given for the human speech. The production of speech is seen as a fourfold process of sound generation, namely *parā, paśyantī, madhyamā* and *vaikharī*. After a brief discussion of modern view of speech production, the Vedic view is discussed.

Speech Production: A Modern View

A question of scientific importance is about the origin of sound. This question, to begin with, can be addressed in reference to human speech. There is a large amount of scientific literature on human speech and music (Honda 2003; Patel 2008). In a paper by Honda (2003), four processes are recognized in production of speech. They are:

(a) Language processing, in which the content of an utterance is converted into phonemic symbols in the brain's language centre;

(b) Generation of motor commands to the vocal organs in the brain's motor centre;

(c) Articulatory movement for the production of speech by the vocal organs based on these motor commands;

(d) Emission of air sent from the lungs in the form of speech.

It is seen from the above description that the emphasis is on neurological and physiological basis. It is the intention of this paper to look into the importance of speech sounds and their production from the Vedic view.

Four Stages of Speech Sounds: Vedic View

An important hymn from *Taittirīya Brāhmaṇa* (Ramabhadracharya 2004; Sharma 2007) says:

catvāri vāk parimitā padāni tāni vidurbrāhmaṇā ye manīṣiṇaḥ |
guhā trīṇi nihitā neṅgayanti turīyaṁ vāco manuṣyāḥ vadanti ||

> Speech or word has four stages, which are known by brāmhaṇa who has control over mind and is well versed in this field. The first three stages are concealed in the cave within while only the fourth stage comes out as utterances by human beings.

According to a view of philosophers of grammar (Sharma 2007) the four stages are referred as (a) *parā*, (b) *paśyantī*, (c) *madhyamā*, and (d) *vaikharī*.

Also in addition to this the scriptures refer to *dhvani* (Mukhopadhyaya 1979), which would mean as voice and is related to speech sounds. The author Śārṅgadeva in *Saṅgīta Ratnākara* (Shringy and Sharma 1978) says:

> *ātmā vivakṣamāṇo 'yaṁ manaḥ prerayate manaḥ* ।
> *dehasthaṁ vahnimāhanti sa prerayati mārutam* ॥
>
> *brahmagranthisthitaḥ so 'tha kramādūrdhvapathe caran* ।
> *nābhihṛtkaṇṭhamūrdhāsyśeṣvāvirbhāvayati dhvanim* ॥

> The human being impels the mind and the mind activates the internal power source in the body, which in turn stimulates the vital force. The vital force stationed around the root of the navel, rising upwards gradually through the heart, and the cerebrum and the cavity of the mouth as it (the vital force) passes through them to manifest as *dhvani*.

In this process also, we can see clearly the four stages for production of sound (*dhvani*) as: (a) the person's impelling of the mind; (b) the mind activating the vital force and getting ready to move; (c) the vital force moving and rising up through the navel, heart, throat, cerebrum and mouth cavity; and (d) *dhvani* produced and travelling to receiver.

Parā, Paśyantī, Madhyamā and Vaikharī

It is to be noted that in the Vedic view the origin of the verbal speech in human beings is traced to the location of the *mūlādhāra cakra*, which is placed in the coccygeal triangle where the spinal cord ends. This energy centre (*cakra*) is of primary importance as it supports the entire body organization (Ramabhadracharya 2003). An important verse from the *Prapañcasāra Tantra* text says:

> *mūlādhārātprathamamudito yastu bhāvaḥ parākhyaḥ* |
> *paścātpaśyantyathā hṛdayāgo buddhiyuṅmadhyamākhyaḥ* ||
>
> *vaktre vaikharyathārurudiṣorasya jantossuṣumnā* |
> *baddhastasmādbhavati pavana prerito varṇa saṅghaḥ* ||
>
> The *parā* form of speech itself was unmanifest to begin with, then it manifested to become *paśyantī* as it moved up with vital force to the region of umbilicus or the navel area. (Also simultaneously, the information at knowledge level is being converted into speech form.) The *paśyantī* form formed at the location of navel region rises up along with the vital force to the location of heart to become *madhyamā* along with the integration of the intellect, to transform knowledge into speech form. Then from *madhyamā* form at the heart it rises along with the vital force to the location of throat and then to mouth to manifest as *vaikharī* speech, or verbal speech heard by the listener. Thus the articulated speech that contains alphabets (and words) has manifested through the movement of vital force through the central energy carrying channel termed as *suṣumnā* along the central axis of the spinal cord.

Thus in the above verse the whole process of fourfold level generation of speech is well described from the Vedic view.

There is a verse from a classical Hindu text, which specifically gives the locations in terms of *cakra*s to the fourfold level of speech in physiological terms in the human body similar to that as the previous verse. The verse says:

> *parāvāṅamūlacakrasthā paśyantī nābhisaṁsthitā* |
> *hṛdisthā madhyamā jñeyā vaikharī kaṇṭhadeśagāḥ* ||
>
> In the fourfold speech, the *parā* level speech manifests at the

> *mūlādhāra cakra* (coccygeal triangle), the *paśyantī* at the *maṇipūra cakra* (navel region), *madhyamā* at the *anāhata cakra* (cardiac or heart region) and the *vaikharī* at the *viśuddhi cakra* (at the cavity of the throat).

It is seen that similar description is given in the *Bhāgavata Purāṇa* (Beck 1993) in the verse 11.12.17. The various *cakras* that are referred in the above verse are shown in *fig*. 7.4.

In another verse, the psychological process of speech formation is given as:

> *vaikharī śabdaniṣpattiḥ madhyamā smṛti gocarā* ।
> *dyotikārthasya paśyantī sūkṣmā brahmaiva kevalam* ॥

> In the fourfold levels of speech, the *vaikharī* form is the gross manifestation of words. The *madhyamā* form is at the memory level, the *paśyantī* form is at the level where the motive is seen and the *parā* form is at the subtle unmanifest as *Brahman*.

The above three important verses clearly show that the Vedic view integrates both the physiological and psychological process involving the human body in explaining the fourfold speech (or sound field) formation through (a) *parā*, (b) *paśyantī*, (c) *madhyamā*, and (d) *vaikharī* levels.

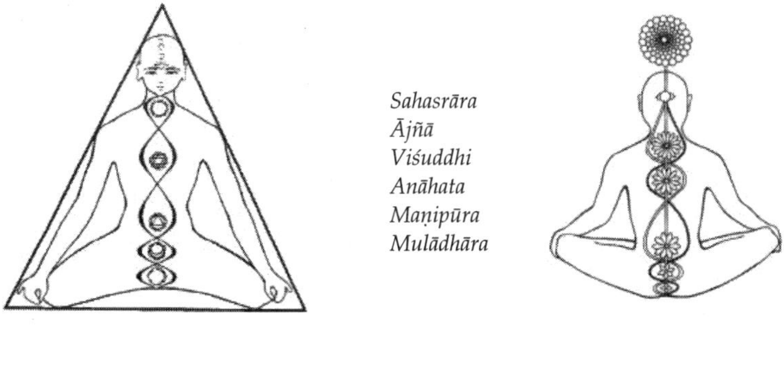

fig. 7.4: Representation of the seven yogic *cakras* in a human being

The *vaikharī* level is the production of words and sentences. This level is physically energized by the vital force (in the form of exhalation and inhalation) at the throat and mouth. Prior to this is the *madhyamā* level, which is drawing from memory (refers to vocabulary and events) and is physically energized by the vital force at the heart or cardiac region.

Prior to this is the *paśyantī* level where the purpose of speech is seen by the speaker and is physically energized by the vital force at the region of navel area or region of umbilicus. Prior to this is the subtlest and unmanifest *parā* level and exists physically as the potential energy form, at the base of the spine or the coccygeal triangle. It is interesting to observe here that in the production of speech and music, not just the contributions of the efforts of the mouth but also of the whole body are involved.

Relevance of the Vedic Fourfold Model of Speech

A comparison of the modern acoustical view and the Vedic view of the production of the speech is hereby considered. It is seen that the modern view considers the role of brain, lungs, and vocal organs, such as throat and mouth. However the Vedic view, through fourfold levels, integrates both the physiological and the psychological aspects. The physiological aspects in the Vedic view go to the base of the spinal cord as the source of physical energy which drives the vital force to rise up and activate the various components such as lungs, heart, throat and the mouth. The psychological aspects do bring up the role of brain through memory and seeing the motive of speech. The Vedic view seems to be more comprehensive. In addition, the Vedic view of speech as a divine energy is important for the human beings in their spiritual development. It is clearly seen that the production of speech involves not just the lungs and mouth but the whole body. It is for this reason that recitation of *mantras* and singing of music are seen as explicit spiritual practices as these affect not only the whole body but also the mind.

The Vedic fourfold level speech production can be applied to any type of sound production. As an example, when one blows a conch shell, the sound produced by shell (heard by the listener) is *vaikharī*. Prior to that the blowing of vibrating air-pulses to the conch shell is the *madhyamā* level. Prior to that of blowing, the person blowing the conch shell had decided to blow for a particular purpose which is seen by the person and is at the *paśyantī* level. Prior to that is the capability or potential of the person to blow a conch shell, which is at the unmanifest level and is referred as *parā*. This Vedic fourfold approach for speech production can be adapted to any production of the sound either through direct vocal music or through an instrument.

Acoustical Aspects of Music

An important role of *vaikharī*, which is the fully manifested sound field, is music. It is well known that music transcends the barriers of language. In the words of Yogī Śrīraṅga Sadguru, "Music should become the bridge that takes the listeners from the sensual level to the spiritual level of *ātman*". The music is universally based on seven notes which are directly related to the accents that are used in the Vedic chants. In Sanskrit the term *svara* denotes a note. *Svara*s refer to accents in the Vedic *mantra*s, notes in music and vowels in language.

The *svara*s in music refer to the basic seven notes found in the seven sounds produced from animals in nature. They are *sā* from a peacock, *re* from a bull, *gā* from a goat, *mā* from a curlew, *pā* from a cuckoo, *dhā* from a horse and *ni* from an elephant. A *rāga* is an acoustic presentation using various *svara*s to depict an emotion or a mood. The various moods are broadly classified into nine moods, namely *śr̥ṅgāra* (amorous), *vīra* (heroic), *karuṇa* (pathetic), *hāsya* (comic), *adbhuta* (marvellous), *raudra* (ferocious), *bhayānaka* (frightening), *bībhatsa* (odious), and *śānta* (peace). The essential effect of the various moods produced by musicians on the listeners is called *rasa*. Thus, it is seen that the acoustical aspects of music have effects on listeners through the tonal configuration, namely *rāga* and total experience, i.e. *rasa* (Rao 2000). Suvarnalata Rao says:

> The different connotations of *rasa* could be viewed as having significance on three different levels — the physical, the psychological and the metaphysical. In the physical sense, it is used to denote juice or essence. For instance, when one states that the fruit is full of *rasa*, it means that it is full of juice. In the psychological sense, it implies flavour or taste, where the active participation of psyche plays a definite role, while on the metaphysical level it describes the experience that one undergoes when aroused by an artistic expression. The perceiver enjoys an experience, which is exalted from the particular to the universal plane by bringing about a complete effacement of the experiencer's ego and its total identification with that artistic creation. It is this supra-mundane experience described in Upaniṣads that assumes significance in Indian aesthetics.

The two types of music are essentially vocal and instrumental. As vocal music comes powerfully through a melodious voice, the instrumental music produced through a well designed and tuned instrument can also be very powerful. Substantial scientific and engineering knowledge and skills are required to build a good instrument. In Indian instrumental music there are hundreds of instruments in use from the ancient times. It is well known that the instruments, namely *vīṇā* (a string instrument) is held by Goddess Sarasvatī and bamboo flute is held by Lord Kr̥ṣṇa as shown in *figs*. 7.5 and 7.6.

fig. 7.5: Goddess Sarasvatī, the Deity of Knowledge, Language and Music

fig. 7.6: Lord Kṛṣṇa with bamboo flute

The divine energy within a human being, which energizes human speech and music, is Goddess Sarasvatī in the form of deity. She is also referred as Vāgdevī or Goddess of Speech. The instrument *vīṇā* she is holding is a string instrument, which has 24 frets. The number 24 also refers to the 24 syllables of the celebrated Gāyatrī *mantra*. In addition, the instrument *vīṇā* also is a representation of the human spinal cord, which also has 24 cartilages. This representation is shown in *fig.* 7.7. Also it is noted in the Yoga and Tantra texts that the locations of various seed-sounds (*bījākṣaras*) is at various *cakras*.

The flute, which is held by Lord Kṛṣṇa, is known for ecstatic music and its spiritual impact on listeners. The epic *Bhāgavata* describes in detail the impact of music played by Lord Kṛṣṇa on the flute. In this context also, the holes in the flute represent the location of the various *cakras*. The measurements of frequencies of various notes produced from a bamboo flute are shown in *fig.* 7.8. Also, the experimental ratios obtained from a bamboo flute compare well with the well-known Pythagorean scale as shown in *fig.* 7.9.

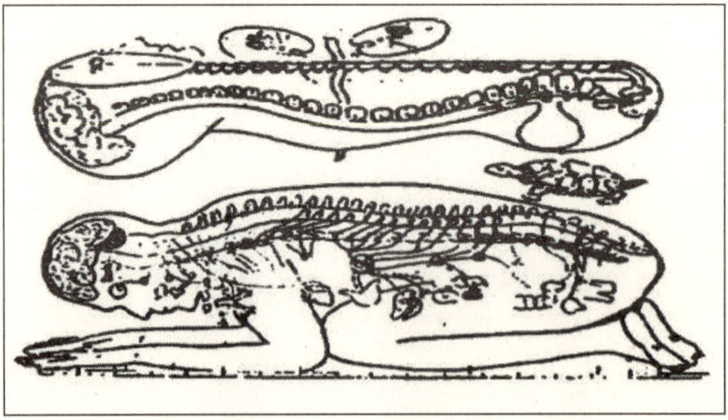

fig. 7.7: Representation of the musical instrument *vīṇā* (above) with spinal cord of a human being (below)

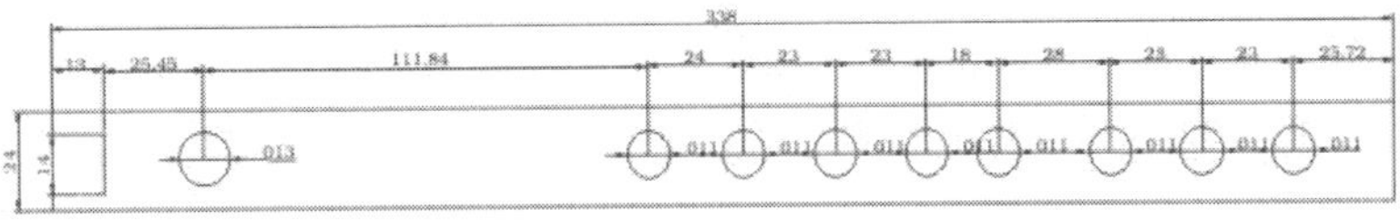

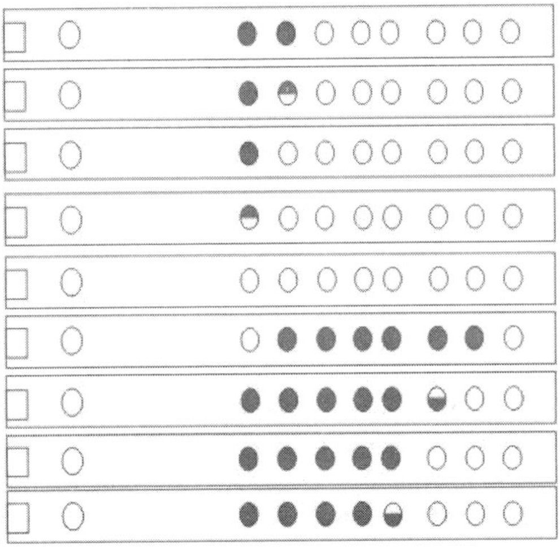

Note	Hz	Expt.	Theory
SA	837	1.00	1.00
RH	910	1.09	1.07
RF	952	1.14	1.13
GH	1030	1.23	1.20
GF	1080	1.29	1.25
MF	1127	1.35	1.33
MH	1227	1.47	1.41
PA	1262	1.51	1.50
DH	1352	1.62	1.60
DF	1380	1.65	1.69
NF	1492	1.78	1.80
NH	1575	1.88	1.88
SA	1660	1.98	2.00

fig. 7.8: Measured frequencies of musical notes from an Indian bamboo flute

NOTE	RATIO (EXPT)	NOTE	RATIO [PYTHOGOREAN]
SA	1.00	C	1.00
RF	1.14	D	1.13
GF	1.29	E	1.25
MF	1.35	F	1.33
PA	1.51	G	1.50
DF	1.65	A	1.66
NH	1.88	B	1.88
SA	2.00	C	2.00

fig. 7.9: Comparison of ratios of measured musical scale from an Indian bamboo flute with the Pythogorean scale

Another interesting feature of the Indian percussion instruments, namely *mṛdaṅga* and *tablā* is that the rhythmic sounds produced from them have melody. This is because their natural modes have harmonic relationships. *Fig*. 7.10 shows the modes and their harmonic relationship. It is well known in musical science that the melody is due to harmonic relationship. It is for this reason that the sounds of *mṛdaṅga* and *tablā* are not only rhythmic but also melodic.

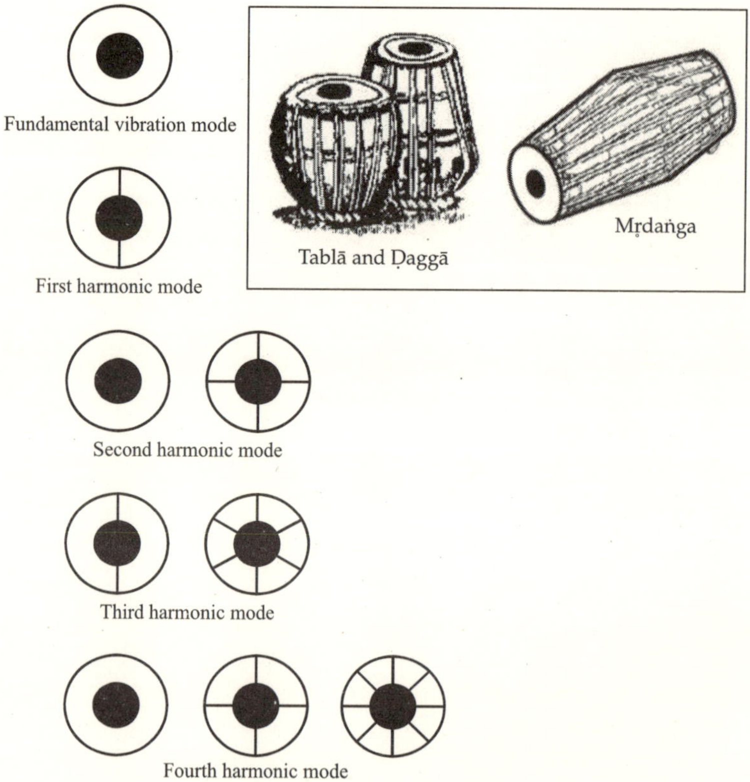

fig. 7.10: Fundamental and four harmonic modes of Indian percussion instruments (*mṛdaṅga* and *tablā*)

In addition to musical instruments, there are additional instruments such as conch shell, bell, cymbals, and singing bowls, etc. which produce sounds that have special effects. They are used in Hindu worship practices to enhance the spiritual experience. For example, the sound from a conch shell has a very sharp resonance, which results in a tonal sound. It is known that tonal sounds help in focusing of the mind.

The *fig.* 7.11 shows the sharp resonance and the dominating first resonance frequency, which indicates the dominating tonal quality of the sound.

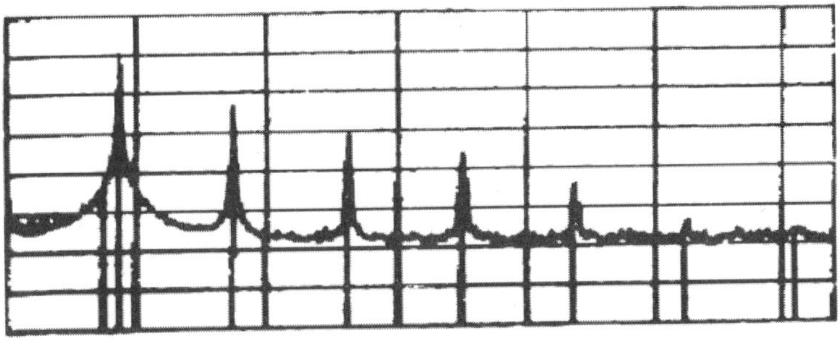

fig. 7.11 (a): Sound spectrum of an Indian conch shell

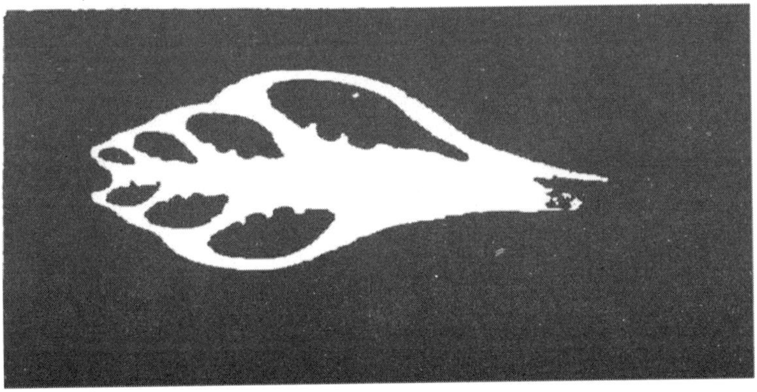

fig. 7.11 (b): X-ray tomography picture of longitudinal section of Indian conch shell

Thus it is seen that the music has significant impact in life. The Vedic literature contains many texts that have dealt in detail about music and its role in life.

Acoustical Aspects of Literature and Language

Literature usually refers to written expressions such as poems, essays, novels, etc. It is well known that poems in particular need to have the acoustical quality when read. The proper choice of words for their sound quality is very important in writing poems. The Sanskrit language, in which most of the Vedic and classical Indian literature is written, has several acoustical features. The distinctive phonetic quality of Sanskrit is reflected in poetry as well as prose. Some of the acoustical aspects of Sanskrit are:

(a) The *sandhi* (joining of two words) has a significant effect on the pronunciation, which in turn has acoustical impact for both speaker as well as listeners (example: *ratha* + *utsava* = *rathotsava*).

(b) The sound corresponding to an alphabet does not change.

(c) The derived words from a verbal root have some acoustic similarity with the sound of the verbal root (example: the root *vid* (to know) becomes *vidvān* (scholar), *vidyārthī* (student), *vidvat* (scholarship), etc.).

(d) The pronunciation of soft aspirant alphabets, hard aspirant alphabets and nasals require different efforts from the speaker and this affects the speaker, and likewise, the listeners are also affected (for example: one needs to pronounce the sound of *ba* softly whereas for producing sound of *bha*, an exhalation force is required).

(e) The large number of alphabets makes it possible to pronounce many complex sounds from the written scripts.

It is well known that the scripts of plays become alive due to the acoustical aspects. The various emotions and their acoustical counterparts are very important in the performances.

Role of Acoustics in Spirituality and Philosophy

Acoustics plays a major role in spiritual aspects of life. As stated earlier, *śabda* and *Nāda-Brahman* refer to the Supreme Being. The five elements of universe, namely earth, water, fire, air and space and their properties are referred in the Vedic literature. The association of the corresponding properties to the element is important in reference to human perception. Several Upaniṣads and scriptures have dealt with this topic (Bahadur 1979). The order of increasing subtlety of the five elements is based on the decreasing number of properties or qualities required to characterize that element. In *Vaiśeṣika-Sūtras* of Kaṇāda Ṛṣi (Bahadur 1979), it is given as:

1. Earth possesses smell, taste, form and touch
2. Water possesses taste, form and touch
3. Fire possesses form and touch
4. Air possesses touch.

The fifth element, termed *ākāśa* (space), cannot be qualified by these four qualities, namely touch, form, taste and smell. This leads to the observation that the space only be characterized or qualified by sound. However, sound is an additional characteristic to qualify all the other four elements, namely air, fire water and earth. In other words, sound as vibration travels through all these four elements and produces different effects, which characterize the corresponding element.

The term *ākāśa* does not refer to emptiness. The term *ākāśa* refers to the substance or element that is spread in all corporeal space meaning that the space capable of containing other elements and bodies (Seal 1985; Watson 1985). It is important to note that the scriptures say that the sound is not the property of air, but air is a carrier of sound waves. This can be further explained using the well-known bell-jar experiment. In this experiment when the air is taken out from the jar to create a vacuum, the listener can see the ringing of the bell, which means that bell surface is vibrating and only the vibrations are not carried further because the air is absent. Changes in air such as turbulence, temperature variations, etc. affect the propagation of sound waves. This means that the air is the efficient cause of sound but not the material cause, because the efficient cause can destroy the property just as the potter's wheel can influence the pot but not the clay (Bharati 1977). This means that sound and space are always together. In the event of an impact the sound generated will propagate through the help of the media such as air, water and earth. Given the existence of space everywhere, it can be inferred that the sound also exists potentially everywhere.

*Mantra*s are used extensively in various spiritual practices. There are three types of practices, namely loud chanting, voiceless chanting and mental chanting. The reason that one should practise in this order is that through loud chanting, one gets the right way of pronunciation and also experiences the acoustical effects. In the course of time, the practitioner will internalize the chanting to the mental level. There also the mystical *bījākṣara*s are used to focus on various deities. The acoustical aspects of these *bījākṣara*s are very important. The spiritual importance of the sound of *OṀ* is well known and has been discussed in several Upaniṣads. Proper collective recitation of chants and *OṀ* result in positive feelings in congregations.

Conclusion

The Vedic fourfold approach to speech sound production provides a deep insight to understanding many phenomena. There are ongoing efforts to analogically view the creation of the universe through this Vedic fourfold view of production of sound field as (a) *parā*, (b) *paśyantī*, (c) *madhyamā*, and (d) *vaikharī* (Dalela 2008). Another way of observation is from the manifest stage to the unmanifest stage, that is to say, from *vaikharī* to *parā*. The interesting feature of the *vaikharī*, *madhyamā*, *paśyantī* and *parā* (VMPP) model is that one can direct the efforts to understand the life and nature, starting from *vaikharī* and proceeding towards *parā*. It is like going from sound to deep silence. The important point to be noted is that there is continuous energy link through vital force from *parā* to *vaikharī*. This means that going in the reverse direction, one should be able to start from *vaikharī* and proceed towards the *parā*, which is spiritually at a very deep level.

Although *vaikharī* is the last quarter, but one can experience *vaikharī* as it is in the manifested form. It encompasses life in many of its expressions such as speech, music, literature and spirituality (Prasad 1998). We know that speech directly relates to human transactions. Music is the human artistic expression (Varma 2009). The literature, such as poetry, plays, comes into life through sound. The spirituality expresses through rituals. *Mantras* are used in meditation practices (Woodroffe 1985; Chamu 2003).

It is known in the spiritual literature that the importance of many types of Yoga are based on the sound field such as *mantra-yoga*, *nāda-yoga*, *japa-yoga*, *svara-yoga* (Brahmavarchas 1995; Swami Sivananda 1994), etc. Thus further work is needed in both theoretical and practical understanding and applications of the sound field (Seal 1985). The fourfold Vedic model for sound field would be helpful for further studies in the field of acoustics (Padoux 1990).

The field of acoustics impacts many aspects of human life. In its desirable form, sound is essential as speech and music. Especially in the form of music and *mantras*, sound plays a major role in improving human life (Prasad 2004). Much research work is further needed on the effect of music and *mantras* from the neuroscience point of view. The sound, in its undesirable form as noise, results in several negative effects on quality of human life (Keizer 2010). Thus, it is important to create an acoustical environment that is conducive to sublime enjoyment of life and achieve spiritual progress.

Acknowledgements

The author is grateful to Yogī His Holiness Sri Rangapriya Swami of Ashtanga Yoga Vijnana Mandiram, Bangalore (India), for his guidance and help. The author also gratefully thanks Dr V. Sadagopan of New York (USA), for his help.

References

Aiyar, K.N., 1997, *Thirty Minor Upaniṣads* (English Translation and Sanskrit Text), Delhi: Parimal Publications.

Bahadur, K.P., 1979, *The Wisdom of Vaisheshika*, New Delhi: Sterling Publishers.

Beck, G.L., 1993, *Sonic Theology*, USA: University of South Carolina Press.

Bharati, Swami Paramananda (T.S. Shankara), 1977, *Reflection and Speculation on Space, Time and Causality*, Madras: Adi Shankara Advaita Centre.

Brahmavarchas, 1995, *Śrī Rāmācārya Vāṅgmay* (Śābda Brahma-Śābda Brahma), Mathura: Akhand Jyoti Samsthan.

Chamu, S.V., 2003, *The Divine Dancer*, Mysore: Ashtanga Yoga Vijnana Mandiram.

Dalela, A., 2008, *Vedic Creationism*, New York, USA: Universe.

Honda, M., 2003, "Human Speech Production Mechanisms", *NIT Technical Review*, **1**(2): 24-29.

Keizer, G., 2010, "The Unwanted Sound of Everything We Want: A Book About Noise", *Public Affairs*, New York, USA.

Mukhopadhyaya, S., 1979, "Importance of Sound in the Tradition of Vedic Chanting", in *Dhvani*, ed. S.C. Malik, New Delhi: Indira Gandhi National Centre of Arts and D.K. Printworld.

Padoux, A. (tr. Gontier, J.), 1990, *Vāk (The Concept of the Word in Selected Hindu Tantras)*, New York, USA: State University of New York Press.

Patel, A.D., 2008, *Music, Language and Brain*, New York, USA: Oxford University Press.

Prasad, M.G., 1998, *Garland: An Anthology of Hinduism*, New York, USA: Association for the Advancement of Arts and Sciences from India.

———, 2004, *Philosophical and Cultural Perspectives of Acoustics in Hinduism*, 147th Meeting of the Acoustical Society of America, New York, USA. (Also the author was invited to write a popular version of this presentation titled "Mantras to Music" which was included in the American Institute of Physics website).

Ramabhadracharya, N.S., 2003, *Śrī Guru Hridaya and Ārṣa-Sāhitya* (Sanskrit text and translations), Mysore: Ashtanga Yoga Vijnana Mandiram.

———, 2004, *Udaka Śānti Mantraḥ* (Sanskrit and translations in Kannada), Mysore: Sri Sanatana Bharati Pathasala.

Raman, C.V., 1922, "The Acoustical Knowledge of the Ancient Hindus", *Āśutoṣa Mukherjī Silver Jubilee Volume*, vol. 2, pp. 179-85.

Rao, Suvarnalata, 2000, *Acoustical Perspective on Raga-Rasa Theory*, New Delhi: Munshiram Manoharlal.

Sadguru, Sri Ranga, 1983–2008, *Amaravāṇī: Lectures on Sanātana Dharma*, Mysore: Ashtanga Yoga Vijnana Mandiram, 19 vols.

Seal, B., 1985, *The Positive Sciences of the Ancient Hindus*, Delhi: Motilal Banarsidass.

Sharma, Rama Nath, 2007, *Vāk, Śabda-tattva and Veda*, Ujjain: World Vedic Conference.

Shivananda, Swami, 1994, *Tantra Yoga, Śābda Yoga and Kriyā Yoga*, Uttar Pradesh: Divine Life Society.

Shringy, R.K. and P.L. Sharma, 1978, *Saṅgīta Ratnākara of Śārṅgadeva* (text and translations), vol. I, Delhi: Motilal Banarsidass.

Varma, R., 2009, *Indian Music*, New Delhi: Sasta Sahitya Mandal.

Watson, I.K., 1985, *Studies in Hinduism*, Delhi: Navarang.

Woodroffe, Sir J., 1985, *Garland of Letters*, Ganesh and Company.

8

Siddhānta-Darpaṇa of Nīlakaṇṭha Somayājī

K. Mahesh and *K. Ramasubramanian*

Introduction

THE Kerala school of Indian astronomy and mathematics (Sarma 1972a, b), that flourished for more than three centuries — starting from Mādhava (CE 1350) of Saṅgamagrāma — is well known for its contributions to mathematics, in particular to that branch that goes by the name of mathematical analysis today (Ramasubramanian and Srinivas 2010). Besides making several important contributions to mathematical analysis which includes discovering the infinite series for sine, cosine and arc-tangent functions as well as its fast convergent approximations,[1] the astronomers of the Kerala school have also made significant contributions to the advancement in astronomy, particularly the planetary theory.

Parameśvara of Vaṭasseri (*c.* 1380–1460) — one of the highly-respected astronomers and disciple of Mādhava (the pioneer of the Kerala school) — is supposed to have made meticulous observations for over a period of fifty-five years continuously. Glowing tributes have been paid by the later astronomers to Parameśvara for this extraordinary feat.[2]

---

1 GYB, chapters 6 & 7.

2 For instance, Nīlakaṇṭha in his *Āryabhaṭīya-Bhāṣya* observes: →

Nīlakaṇṭha Somayājī of Trikkaṇṭiyur (*c.* 1444–1550), a disciple of Parameśvara's son Dāmodara, and the author of *Siddhānta-Darpaṇa* (the work currently being studied), carried out a fundamental revision of the traditional planetary theory (Ramasubramanian, Srinivas and Sriram 1994). In the second chapter of his treatise *Tantrasaṁgraha*,[3] Nīlakaṇṭha presents a major revision of the earlier Indian planetary theory for the interior planets Mercury and Venus. Based on an incisive analysis of the period of latitudinal motion of the interior planets, Nīlakaṇṭha came up with a much better formulation of the equation of centre of these planets than was available either in the earler Indian works or in the Islamic or the Greco-European traditions of astronomy till the work of Kepler — that was to come more than 100 years later. This also, incidentally, led Nīlakaṇṭha to give a unified formulation for the procedure for the computation of latitudes (Sriram 2002).

It may also be mentioned that Nīlakaṇṭha was the first savant in the history of astronomy to clearly deduce purely from the observed motion of the planets — and not from any speculative or cosmological arguments — that the interior planets go around the Sun and that the period of their motion around the Sun is also the period of their latitudinal motion. He explicitly mentions this in his *Āryabhaṭīya-Bhāṣya*:[4]

तयोः भ्रमणवृत्तेन न भूः कबलीक्रियते।

tayoḥ bhramaṇavr̥ttena na bhūḥ kabalīkriyate ।

→ परमेश्वराचार्येण पुनः ग्रहणग्रहयोगादिकं यन्त्रैः पञ्चपञ्चाशद्वर्षकालं सम्यक् परीक्षितम्। . . . तस्य गोलवित्तमत्वं च गोलदीपिकादिभिः तत्कृतैः प्रबन्धविशेषैः ज्ञायते। अत अन्येषां करणानां स्वस्वकाले यावत्सूक्ष्मत्वं ततोऽप्येतत्कृतस्य दृग्गणिताख्यस्य सूक्ष्मत्वमुपद्यते। अपि च परमेश्वरोक्तमन्दोच्चपरिध्यादीनाम् आर्यभटोक्तेभ्यः सूक्ष्मत्वमिदानीमपि ग्रहयोगादिष्वस्माभिर्ज्ञायते। (*ABB*, p. 154).

3 *TS*, chapter 2.

4 *ABB*, p. 9.

> The Earth is not circumscribed by the orbit of them (Mercury and Venus).

In his other works, which includes *Golasāra* and *Siddhānta-Darpaṇa*, Nīlakaṇṭha describes the geometrical picture of planetary motion that follows from his revised model. In a remarkable short tract called *Grahasphuṭānayane vikṣepavāsanā*,[5] which forms a part of the text called *Gaṇitayuktayaḥ*, Nīlakaṇṭha succinctly yet definitively describes his cosmological model.[6] As per the model proposed by Nīlakaṇṭha, the five planets Mercury, Venus, Mars, Jupiter and Saturn move in eccentric orbits which are inclined to the plane of ecliptic around the mean Sun, which in turn goes around the Earth (Ramasubramanian, Srinivas and Sriram 1994).

Nīlakaṇṭha and His Works

Nīlakaṇṭha is one of the very few astronomers about whom we are able to gather a fairly detailed information regarding the date and place of birth, his ancestors, family deity, the siblings he had, the *gurus* from whom he received his teaching on different disciplines and so on. This is all available to us from his own writings as well as citations made by others.

FAMILY DETAILS OF NĪLAKAṆṬHA

One of the most important sources of information regarding the personal details of Nīlakaṇṭha is the colophon that has been left behind by him at the end of Gaṇitapāda of his *Āryabhaṭīya-Bhāṣya*. The long sentence supplying this information reads as follows:

> इति श्री-कुण्डग्रामजेन गार्ग्यगोत्रेण आश्वलायनेन भाट्टेन केरलसद्ग्रामगृहस्थेन श्रीश्वेतारण्यनाथ-परमेश्वरकरुणाधिकरणभूतविग्रहेण

[5] This short track seems to have been written after authoring the monumental treatise *Āryabhaṭīya-Bhāṣya* as one finds the citation from the latter in the former.

[6] *GVV*.

जातवेदःपुत्रेण शङ्कराग्रजेन जातवेदोमातुलेन दृग्गणितनिर्मापकपरमेश्वरपुत्र-श्रीदामोदरात्तज्योतिषामयनेन रवित आत्तवेदान्तशास्त्रेण सुब्रह्मण्यसहृदयेन नीलकण्ठेन सोमसुता विरचितविविधग्रन्थेन दृष्टबहूपपत्तिना स्थापितपरमार्थेन कालेन शङ्कराय निर्मिते श्रीमदार्यभटसिद्धान्तव्याख्याने महाभाष्ये . . .

iti śrī-kuṇḍagrāmajena gārgyagotreṇa āśvalāyanena bhāṭṭena keralasadgrāmagr̥hasthena śrīśvetāraṇyanātha-parameśvara-karūṇādhikaraṇabhūtavigraheṇa jātavedaḥputreṇa śaṅkarā-grajena jātavedomātulena dr̥ggaṇitanirmāpakaparameśvara-putraśrīdāmo-darāttajyotiṣāmayanena ravita āttavedānta-śāstreṇa subrahmaṇyasahr̥dayena nīlakaṇṭhena somasutā viracitavividhagranthena dr̥ṣṭabahūpapattinā sthāpita-paramārthena kālena śaṅkarāya nirmite śrīmadārya-bhaṭa-siddhāntavyākhyāne mahābhāṣye . . .

GURUS OF NĪLAKAṆṬHA

In the passage cited above, Nīlakaṇṭha has clearly mentioned that he learned Jyotiḥśāstra from Dāmodara (and Vedānta from Ravi). This in fact, gets re-affirmed from the invocatory verse that we find in the *Siddhānta-Darpaṇa.*

श्रीमद्दामोदरं नत्वा भगवन्तं रविं तथा ।
तत्प्रसादान्मया लब्धं ज्योतिश्चरितमुच्यते ।।

śrīmaddāmodaraṁ natvā bhagavantaṁ raviṁ tathā।
tatprasādānmayā labdhaṁ jyotiścaritamucyate।।

The nature of [motion of] the celestial objects (*jyotiścaritam*) is being narrated by me, having offered my venerations to Dāmodara and likewise to Lord Ravi, through whose grace it was received by me.

Given the fact that it is customary in the Indian tradition to offer veneration to the Lord as well as one's own teacher before commencing any purposeful undertaking (particularly the auspicious ones), and the fact that in the latter half of the verse it is mentioned that the author gained the knowledge of Jyotiḥśāstra through the grace of Dāmodara, it gets easily established that Dāmodara was Nīlakaṇṭha's *guru* for Jyotiḥśāstra.

WORKS OF NĪLAKAṆṬHA

Through his scholarly investigations, K.V. Sarma has identified nine works to have been composed by Nīlakaṇṭha. Unfortunately, manuscripts of only six of them (see Table 8.1) have been discovered so far,[7] and luckily all the six of them are available in printed form.

Table 8.1: The Published Works of Nīlakaṇṭha

Name of the Work	*Brief Description*	*Style*
Tantrasaṁgraha	Comprehensive Text on Indian Astronomy	Poetry
Golasāra	Essence of Spherical Astronomy	Poetry
Candracchāyā-gaṇita and its *Vyākhyā*	Computations Related to Moon's Shadow	Poetry and prose
Siddhānta-Darpaṇa and its *Vyākhyā*	Mirror of the Principles of Astronomy	Poetry and prose
Jyotirmīmāṁsā	Investigations on the Principles of Astronomy	Prose
Āryabhaṭīya-Bhāṣya	An Elaborate Commentary on *Āryabhaṭīya*	Prose

Of the six works listed in Table 8.1, the first two are purely in the form of metrical composition. While a couple of others (5 and 6) are purely in the form of prose, the other two comprise both prose and metrical composition.

Among the works listed in Table 8.1, the only text that has been studied *in toto* so far is *Tantrasaṁgraha*.[8] Studies related to the other texts (2 to 6, of which the 6th is published in three

---

[7] The three works yet to be traced are: *Grahaṇanirṇaya, Sundararāja-Praśnottara* and *Graha-parīkṣākrama*.

[8] Recently, a translation of *Tantrasaṁgraha* (item 1) with mathematical notes and several useful appendices has been published (see *TS*).

volumes) are either partially done[9] or are yet to be taken up.

Particularly, the motivation to make a critical study of *Siddhānta-Darpaṇa*[10] stems from the following consideration. If we were to list the various compositions of Nīlakaṇṭha in a chronological order, it seems *Siddhānta-Darpaṇa* and its *Vyākhyā* would be somewhere in the tail of the list.[11] Hence, one expects the text to mirror Nīlakaṇṭha's ripe understanding of the subject matter in general and towards the planetary theory in particular — to the advancement of which his contribution has been remarkable as mentioned in the introduction.

Siddhānta-Darpaṇa and its Contents

Siddhānta-Darpaṇa is a short treatise consisting of only thirty-two verses. However, the information that has been packed in this text is quite enormous. Nīlakaṇṭha, following the style of *Āryabhaṭīya*, has composed some of the verses in a highly compressed and cryptic form, perhaps for easy oral onward transmission from the teacher to the taught.

True to the title, the text *Siddhānta-Darpaṇa* (Mirror of the Principles of Astronomy)[12] not only presents a set of parameters and lays down the procedures required for the computation of planetary longitudes, but also discusses the principles based on which such procedures have been enunciated by the ancients. The material presented here is well organized and has been developed in a very structured manner. Leaving out the first (invocatory) and the last (concluding) verses, the text

---

9 Only sections of these texts have been studied and cited. A full translation and notes are yet to be done.

10 *SD*.

11 That it has been composed at a ripe old age is evident from the fact that he cites *Āryabhaṭīya-Bhāṣya* in *Siddhānta-Darpaṇa-vyākhyā:*

एतत्सर्वं मया आर्यभटीयव्याख्याने प्रपञ्चितमिति विरम्यते। (*SD*, p. 24)

12 *siddhāntānām darpaṇaḥ = siddhāntadarpaṇaḥ* ।

consists of only thirty verses. This has been divided into two parts: the Upadeśa-bhāga (twenty verses) and the Nyāya-bhāga (ten verses). In fact, this division has been indicated by Nīlakaṇṭha himself in the last verse of the text which goes as:

> विंशत्यनुष्टुभां स्पष्टं कृतं शास्त्रमिहाखिलम् ।
> दशभिर्न्यायभागश्च संक्षेपादेव दर्शितः ।।
>
> *viṁśatyanuṣṭubhāṁ spaṣṭaṁ kr̥taṁ śāstramihākhilam* ।
> *daśabhirnyāyabhāgaśca saṁkṣepādeva darśitaḥ* ।।
>
> While, the entire science [needed for computational purposes] has been delineated in twenty verses in *anuṣṭubh* metre, the theory backing it up has been explained very briefly in ten verses.

The usage of the phrase *śāstramihākhilam* (here, the entire science) in the above verse merits our attention. The word "entire" should not be taken in its expressed sense. Though it has been employed in a restricted sense[13] confining the boundary to the computation of planetary positions, the usage is justified in the sense that it is the computation of longitudes that forms the basis for every other computation — be it *lagna*, latitude, eclipse, etc., that is done in astronomy. In what follows we briefly outline the contents of the two parts.

PART 1: UPADEŚA-BHĀGA

The word *upadeśa* literally means instruction or advice. True to the name, in this part (Table 8.2) of *Siddhānta-Darpaṇa*, Nīlakaṇṭha presents all the basic astronomical constants — perhaps as verified and adopted by him — such as the number of revolutions made by the planets including their aphelions and nodes, that would be useful in the computations of planetary longitudes, latitudes, eclipses and so on. Incidentally,

---

13 In technical Sanskrit such a process of restraining the meaning of a word is called *saṅkoca*.

these parameters given are for a very large period of time known as *kalpa*. A *kalpa* consists of 1,000 *mahāyugas* which correspond to a gigantic figure of 4.32×10^9 years. Along with the revolution number of planets and their aphelion, Nīlakaṇṭha also specifies the dimensions of the *manda* and *śīghra* epicycles, that are involved in transforming the mean longitudes[14] (*madhyamagraha*) of the planets into the observed geocentric longitudes (*sphuṭagraha*). There is also a brief mention of the motion of equinoxes in the text, and an elaborate commentary on it. The part concludes with a succinct, yet definitive description of the geometrical picture of planetary motion implied by the computational scheme.

Table 8.2: The Contents of Upadeśa-bhāga

Verse(s)	*Content of the Verse(s)*
1	मङ्गलाचरणम् (*maṅgalācaraṇam*) (Invocation)
2-5	ग्रहपर्यया: (*grahaparyayāḥ*) (Revolutions made by the planets in *kalpa*)
6-8a	मन्दोच्चपातपर्यया: (*mandoccapātaparyayāḥ*) (Revolutions of *mandoccas* and *pātas*)
8b	परमविक्षेपकला: (*paramavikṣepakalāḥ*) (Inclination of the planetary orbits)
9-10	मन्दपरिधय: शीघ्रपरिधयश्च (*mandaparidhayaḥ śīghraparidhayaśca*) (Circumferences of *manda* and *śīghra* epicycles)
11-12	युगमानम् (*yugamānam*) (Measure of aeons)
13a	भगणांशा: (*bhagaṇāṁśāḥ*) (Units of circle)
13b	ग्रहयोजनभुक्ति: (*grahayojanabhuktiḥ*) (Linear distance covered by the planets (in *yojanas*))
14	भूमीन्दुरविव्यासा: (*bhūmīnduravivyāsāḥ*) (Diameters of the Earth, Moon and the Sun)

Cont.

14 The mean longitude of the planets is obtained by assuming that they are moving uniformly on the *kakṣyāmaṇḍala* (deferent circle).

Verse(s)	*Content of the Verse(s)*
15a	अवन्तीस्थानम् (*avantīsthānam*) (Location of the city Avantī)
15b	नक्षत्रपरिधिः (*nakṣatraparidhiḥ*)(The perimeter of the apparent orbit of the stars)
16	अपक्रमवृत्तस्थितिः (*apakramavṛttasthitiḥ*) (The position of the ecliptic)
17	अयनचलनम् (*ayanacalanam*) (Motion of equinoxes)
18	विदितकाले अयनस्थितिः (*viditakāle ayanasthitiḥ*) (Position of equinoxes on a specific date)
19-21	ग्रहभ्रमणप्रकारः (*grahabhramaṇaprakāraḥ*) (Geometrical picture of planetary motion)

PART 2: NYĀYA-BHĀGA

The second part (Table 8.3) of *Siddhānta-Darpaṇa* as the title itself suggests discusses the rationale behind some of the mathematical and astronomical procedures described in the earlier part. Particularly, here Nīlakaṇṭha sets forth how the eccentric and concentric circles (*pratimaṇḍala* and *kakṣyāmaṇḍala*),

Table 8.3: The Contents of Nyāya-bhāga

Verse(s)	*Content of the verse(s)*
22-24	प्रतिमण्डलं कक्ष्यामण्डलं च (*pratimaṇḍalaṁ kakṣyāmaṇḍalaṁ ca*) (Eccentric and orbital circles)
26	भुजादिलक्षणम् (*bhujādilakṣaṇam*) (Definition of Sine, etc.)
27	ग्रहस्फुटः (*grahasphutaḥ*) (Obtaining the true position of a planet)
28-29	क्रान्तिः (*krāntiḥ*) (Declination (of the *manda-vṛttas*))
30a	व्यतीपातः (*vyatīpātaḥ*) (*Vyatīpāta*: *Lāṭa* and *Vaidhṛta*)
30b	द्विस्पृग्वृतम् (*dvispṛgvṛtam*) (The circle passing through centre of the Sun and the Moon)
31a	ग्रहणादिः (*grahaṇādiḥ*) (Eclipses, etc.)
32	ग्रन्थसमाप्तिः (*grantha-samāptiḥ*) (Conclusion)

together explain the different types of motion executed by the planets as observed by an observer on the Earth, who is removed from the centre of motion of the planet. He also explains the equivalence of the eccentric and the epicyclic models of planetary motion. Following this, Nīlakaṇṭha succinctly points out the special features that need to be considered while describing the motion of the inner planets as opposed to that of the exterior ones.

Vyākhyā of Siddhānta-Darpaṇa: Its Style, Scope and Special Features

A special feature of the text *Siddhānta-Darpaṇa* is that it has an auto-commentary. It is quite uncommon to find texts of this nature, wherein the author himself has ventured to write an elaborate commentary clarifying various issues and the doubts that are likely to arise in the minds of the reader as they go through some of the terse verses. In fact, one of the hallmarks of Nīlakaṇṭha's commentaries — whether it is *Āryabhaṭīya-Bhāṣya* or *Siddhānta-Darpaṇa-Vyākhyā* or any other work for that matter — is that, he often transcends the confines of immediate utility of the commentary and discusses issues of larger interest. These discussions contribute a great deal to a better understanding of the subject and enable us to appreciate deeper connections with the other branches of knowledge.

Some of his general observations regarding the nature of science, mathematical proof and the like are also very insightful and thought-provoking. To prove the point as it were, in what follows, we briefly recall his observations in connection with (i) the specification of number of rotations made by the earth in a *kalpa*, (ii) the discussion on *ayana-calana*, and (iii) the dissectional proof of *bhujā-koṭi-karṇa-nyāya*.

RECONCILIATION OF DIFFERENT VIEWS

While specifying the number of sidereal days in a *kalpa*

(4.32×10^9), which is the same as the number of star rises; Nīlakaṇṭha presents this as the number of revolutions of the Earth. In his own words:

> .... क्षिते:॥2॥
> खखेषुगोगुणाष्टाश्वाग्न्यश्विद्व्यष्टशरेन्दव:॥3a॥
>
> .... *kṣiteḥ* ॥2॥
> *khakheṣugoguṇāṣṭāśvāgnyaśvidvyaṣṭaśarendavaḥ* ॥3a॥

> . . . . 15,82,23,78,39,500 is [the number of revolutions] of the Earth.

What is interesting to note here is the use of the word *kṣiteḥ*, (which means "of the earth") while specifying the number 15,82,23,78,39,500.[15] In other words, what we want to point out is, Nīlakaṇṭha instead of associating this number with the number of revolutions made by starry sphere has associated it with the Earth. That this association is quite consciously and purposefully done becomes quite evident from the commentary. However, Nīlakaṇṭha has not argued at great length to prove that it is only that the Earth and not the starry sphere that rotates. This can be appreciated considering the following fact. The weight of tradition in any scientific discipline cannot be overemphasized, and it is only to be anticipated that there would be tremendous opposition to forsaking any established viewpoint and adopting a new one. History of science tells us that this would be true even when new evidences clearly indicate the non-viability of old views, let alone new formulations based only on theoretical arguments. Under such circumstances, it would be wise to be

---

15 The ratio of this number to the revolutions of the Sun (which is nothing but the number of years) turns out to be

$$\frac{15,82,23,78,39,500}{432,00,00,000} = 366.25876.$$

This number diminished by one gives the duration of a sidereal year. (The currently accepted value is 365.256363004 days.)

pragmatic, and this seems to be the course adopted by Nīlakaṇṭha, when he leaves the discussion by saying that the two pictures are equivalent.

. . . उभयथापि ग्रहकर्मणि न विशेष इत्यभिप्रायः।

. . . *ubhayathāpi grahakarmaṇi na viśeṣa ityabhiprāyaḥ* ।

Both the ways, there is no distinction in the planetary calculations.

THE NEED FOR OBSERVATION

There is a general view or rather a misconception that in the Indian "philosophical" tradition all *pramāṇa*s are taken to be subservient to *śabda* (and some try to extend it to Indian "scientific" tradition as well) which is not quite correct. In fact, such views far from being true are also misleading.

We mention this here because, Nīlakaṇṭha in his *Siddhānta-Darpaṇa* as well as other work such as *Jyotirmīmāṁsā* had categorically stated that as far as Jyotiṣaśāstra is concerned (which is applicable to other Śāstras as well) all *pramāṇa*s are subservient to *pratyakṣa*. Though the data recorded, based on past observations, plays a crucial role in a science like astronomy[16] — conducting experiments and analysing the data collected is indispensable. In other words, *pratyakṣa* occupies a higher place in hierarchy of *pramāṇa*s. In a field like astronomy any *pramāṇa* in contradiction to *pratyakṣa* would be given a back-seat. Explaining this point in connection with *ayanacalana*, Nīlakaṇṭha observes:

अतः सिद्धं परीक्षणेनैव तद्व्यक्तिः इति।

ataḥ siddhaṁ parīkṣaṇenaiva tadvyaktiḥ iti ।

---

16 For instance for predicting accurate value of *ayanāṁśa*, we need data collected by previous generations. This is because even life-time observation can hardly help arriving at reasonable values because the motion is so slow ($\approx 50''$ per annum).

> Thus, its position is only [is to be determined] only by careful observation.

In the same vein, in yet another context Nīlakaṇṭha points out that the *vikṣepa* (celestial latitude) of stars beyond 5 or 6° on either side of the ecliptic to be observed and determined by means of instruments, for they cannot be gauged in relation to the planets, all of which lie within the said limit.

> अपक्रममण्डलासन्नानि पञ्चषभागविक्षेपान्येव शुक्रादिभिः छाद्यानि। ततो विप्रकृष्टानां विक्षेपाः योगभागाश्च यन्त्रैरेव आनेयाः।
>
> *apakramamaṇḍalāsannāni pañcaṣabhāgavikṣepānyeva śukrādibhiḥ chādyāni। tato viprakṛṣṭānāṁ vikṣepāḥ yogabhāgāśca yantraireva āneyāḥ।*
>
> Only those [stars] which are near to the ecliptic and having latitudes [within] five to six degrees will be covered by the Venus, etc. The latitudes and the degrees of conjunction of others which are deviated more than that can be calculated only by using instruments.

MEANS OF UNDERSTANDING THE VALIDITY OF A RESULT

In the course of his discussion on *bhujā-koṭi-karṇa-nyāya*, considering the specific problem of proving the validity of *bhujā-koṭi-karṇa-nyāya*, Nīlakaṇṭha points out that there are two different approaches that can be taken to demostrate the result:

1. start listing all the triplets (a, b, c) that satisfy the relation $\sqrt{a^2 + b^2} = c$ (where all the three a, b and c are integers), or
2. resort to the geometrical approach, which pervades every domain [of knowledge] (*aśeṣaviṣayavyāpī*).

Having mentioned the two approaches, Nīlakaṇṭha quickly puts forth his reasoning as to why one should first resort to the latter approach than the former. He says that in the case of the former, the domain of applicability is limited, the set will be enormous in size and will be never-ending. What is noteworthy is the choice of Nīlakaṇṭha, of the geometrical method to the enumerative one. Even if the number of observations of the particular instances, could be very large, it is not going to be complete (ending). The word *viśrānti* means taking rest. In this context, it refers to resting after listing sets of three numbers ("Pythagorean" triplets). That the list of elements (set of three numbers), in this set is not finite, is indicated by the use of the word *aviśrānti* (never-ending).

Towards the end of his discussion, having provided the dissectional proof of the *bhujā-koṭi-karṇa-nyāya*, Nīlakaṇṭha makes an interesting comment that stresses the need for skilful reasoning in trying to understand the truth of a given mathematical result.

एतत्सर्वं युक्तिमूलमेव, न त्वागममूलम्।

etatsarvaṁ yuktimūlameva, na tvāgamamūlam ।

> All this is based on reasoning alone and does not require the authority of any authentic text [for its validity].

When analysed carefully, this statement by Nīlakaṇṭha clearly mirrors the underpinning thought process regarding the process of learning as conceived by him.

Generally people tend to think that learning takes place by passive transmission of knowledge from one to the other. This transmission of information may be a necessary part of learning, but not sufficient. Ongoing research by several groups, increasingly point to the fact that people learn the best not by "passively" receiving information (*āgamamūla*) but by actively engaging in thinking and exploring (*yuktimūla*) for themselves.[17] This message has been conveyed in the form of a maxim: *etatsarvaṁ yuktimūlameva, na tvāgamamūlam*.

[17] Today more and more schools are focusing on "learning by doing" and not merely by listening or reading.

Details of the Manuscript

As per the information provided by K.V. Sarma, the text of *Siddhānta-Darpaṇa* has been prepared by consulting seven manuscripts, all of which are in the Malayalam script on palm leaf.

As far as the *Vyākhyā* is concerned, it has been stated that Sarma could find only one manuscript. It has also been noted that the manuscript originally belonged to H.H. The Maharaja's Palace Library, Trivandrum, and is currently preserved at the Kerala University's Oriental Manuscript Library. Unfortunately, certain portions of the manuscript have been lost which include commentary of the verses 7c-16 and 28-32. By a streak of luck if one could trace any other manuscript containing the commentary of this section, it would indeed be very useful as it is here Nīlakaṇṭha discusses some important topics like declination and eclipses.

Appendices Added to the Text

In his edition of the text *Siddhānta-Darpaṇa*, K.V. Sarma has included two appendices:

Appendix I: *Siddhānta-darpaṇasiddha-paryayādayaḥ*

Appendix II: *Siddhānta-darpaṇastha-paryaya-bhūdināni*

In his appraisal about the nature of these appendices, it has been mentioned by Sarma that these appendices are later anonymous compilations. Barring the details, both the appendices essentially present the parameters involved in the computation of planetary positions in *kaṭapayādi* notation. Considering the fact that the parameter values have been specified in the text using *bhūtasaṁkhyā* system, it seems that one of the purposes for which these appendices might have been composed is to present the set of crucial parameters involved in the planetary computation in the more familiar and unambiguous *kaṭapayādi* notation of expressing numbers.[18]

Appendix I contains seventeen verses whereas Appendix II contains only three verses. In the latter, only the revolutions of the planets — including that of the apsis and the node of the Moon — and the number of mean solar days in a *kalpa* are specified. Whereas Appendix I includes certain other useful information such as, the initial positions of the planets at the beginning of the Kali era, the positions of the higher apsis (*ucca*) and the nodes of the planets in the Kali year 4800 (CE 1698), the position of the planets on a specific Kali day (in CE 1704), etc. as calculated with the constants enunciated in the *Siddhānta-Darpaṇa* specified.

Brief Summary of the Analysis

From a careful study of the *Siddhānta-Darpaṇa* along with its *Vyākhyā* (an auto-commentary), it is quite evident to us that Nīlakaṇṭha's contribution as a researcher is outstanding to the advancement in planetary theory. Both as an author of independent works as well as a commentator, Nīlakaṇṭha has done a remarkable job in taking the tradition forward by leaps and bounds. The wealth of quotations found in his commentaries — particularly in *Āryabhaṭīya-Bhāṣya* and in *Siddhānta-Darpaṇa-Vyākhyā* — and the pithy comments he makes either at the beginning of a discussion or towards the end of it, reflect his deep scholarship not only in the science of astronomy, but also in a variety of other disciplines as well. His innate ability to present even the difficult conceptions in a lucid manner is unparalleled. In general, his writings are marked with brevity and exceptional clarity. So much so, there is no overstating in K.V. Sarma describing Nīlakaṇṭha as "one of the brightest stars in the firmament of Indian astronomy" (Sriram, Ramasubramanian and Srinivas 2002).

---

18 Nīlakaṇṭha has chosen *bhūtasaṁkhyā* system to represent numbers in text *Siddhānta-Darpaṇa*.

Among the several works in Indian astronomy, *Siddhānta-Darpaṇa* stands out in various respects, some of which are highlighted below:

1. While most of the texts simply give the procedure for the computation of planetary positions, *lagna*, eclipse, etc. in this particular text focus has been on providing the right parameters and explaining the rationale behind the planetary computational scheme and not merely delineating the procedures which can be found in any siddhāntic work. In this connection he also elucidates the principles involved behind the interactive procedure in computing the *aviśiṣṭa-mandakarṇa*.
2. The picture of planetary motion described in the three verses (19-21) by Nīlakaṇṭha is yet another remarkable feature of the text. Such a description, which is at once clear and brief, is not to be found in the works of other earlier astronomers other than Parameśvara — the *paramaguru* of Nīlakaṇṭha. Though Parameśvara's description in his *Bhaṭadīpikā* of *Mahābhāskarīya* is marked with clarity, it does not score in brevity.
3. In the context of specifying the number of revolutions made by the planets in a larger period *kalpa* (4.32×10^9 years), having stated a huge number (15,82,23,27,-78,39,500, which corresponds to the number of sidereal days in a *kalpa*), Nīlakaṇṭha associates this with the number of rotations of the Earth (*kṣiteḥ*). What is interesting here is not only the revival of a forsaken view of the rotation of the Earth as proposed by *Āryabhaṭīya*, but also the justification presented to show that both the views are equivalent.
4. One also finds a detailed description of the motion of equinoxes (*ayana-calana*) in the commentary which is hardly discussed by other astronomers.

5. While discussing *ayana-calana*, Nīlakaṇṭha mentions that the position of equinoxes was precisely determined by his *paramaguru* Parameśvara through his own meticulous observation.[19] Then giving the approximate position of equinoxes in a particular year, he also specifies an *ahargaṇa* (1660181), and adds that this particular choice was done in order to indicate his birth time (*svajanmakāla-jñāpanārtham*). Inclusion of such biographical details is very rare to find in other texts which provides an indisputable evidence for fixing the date of Nīlakaṇṭha.

6. Before commencing his discussion on the rationale behind a particular correction procedure, Nīlakaṇṭha generally presents the domain of applicability of that correction. In one instance Nīlakaṇṭha imagines a moon-centric observer trying to find out the rate of motion of the Sun. Though such a situation is impractical, that Nīlakaṇṭha could think of such an observer not only forms a shining example to indicate the richness of his imagination, but also shows that Nīlakaṇṭha was far ahead of his times!

7. A graphic description on the dissectional proof of the *bhujā-koṭi-karṇa-nyāya* and some of the epistemological issues discussed therein are not to be found in any other text on astronomy or mathematics. Similarly, a long discussion on the validity of the application of *trairāśika*, when the quantities in *trairāśika* are continuously changing with time, seems to be unique to this text.

8. In yet another context, Nīlakaṇṭha poses a question: where should the observer be located for him to see

---

19 भार्गव-परमेश्वराचार्येण अस्मत्परमगुरुणा चलांशस्त्वम् इति कल्यब्दे परीक्ष्य पञ्चदशांशपूर्तिर्निर्णीता।

the full phase of the Moon all the time? While answering this Nīlakaṇṭha introduces the word *dvispṛgvṛtta* (a circle that passes through two objects) and mentions that the observer has to be at its centre.

Another remarkable feature of *Siddhānta-Darpaṇa* lies in lifting a corner[20] of the cloak of darkness that has been hiding the geometrical picture of the planetary motion behind the computational procedure. In his commentary to verses 19-22, Nīlakaṇṭha explains elaborately the geometrical picture of planetary motion underlying the computational scheme followed by astronomers for several centuries before him. Also, the contribution of Nīlakaṇṭha can be better appreciated by travelling back in time, a time during which the community of astronomers did not have sophisticated astronomical instruments to verify their hypothesis and demonstrate it to the world. In fact, the discussion on several conceptions presented by Nīlakaṇṭha — particularly when he presents his analysis on *ayana-calana* or the picture of planetary motion — gives us a glimpse as to how people of genius, like Āryabhaṭa, Brahmagupta, Bhāskara, Parameśvara and others, stumbled and groped their way through the creative process to arrive piecemeal at their results.

In the context of discussing the *bhujā-koṭi-karṇa-nyāya*, Nīlakaṇṭha beautifully articulates the spirit of scientist. He says:

इत्येतत्सर्वं द्रव्यभूतस्य पुरुषस्य अवश्यं स्फुरति, तस्य संवेद्यसम्बन्धि-अखिलन्यायावयवानां तत्त्वज्ञानपर्यन्तं जिज्ञासाया अविश्रान्तेः।

ityetatsarvaṁ dravyabhūtasya puruṣasya avaśyaṁ sphurati, tasya saṁvedyasambandhiakhilanyāyāvayavānāṁ tattva-jñānaparyantaṁ jijñāsāyā aviśrānteḥ ।

All these become evident to any serious (seriously thinking) person, since his urge to know (*jijñāsā*) never subsides till

---

20 A corner, since not all aspects of astronomy have been covered by him.

he clearly understands all the principles involved in the phenomenon he wants to understand.

Though this is a passing remark that Nīlakaṇṭha makes, it is quite profound in its content. The idea behind this statement is perhaps to get across the message that the urge to understand the fundamental principles "thoroughly" can never be allowed to subside — a hallmark that distinguishes a true scientist from others.

In the following section we present an interesting dissectional proof of the so-called Pythagorean Theorem (the most fundamental theorem of geometry) as presented by Nīlakaṇṭha to just give a flavour of the text *Siddhānta-Darpaṇa* and its auto-commentary by him.

Dissectional Proof of Bhujā-Koṭi-Karṇa-Nyāya

दोः कोटिभुजयोर्योगः कर्णबाहुचतुर्भुजः॥२५॥

doḥ koṭibhujayoryogaḥ karṇabāhucaturbhujaḥ॥25॥

The sum of the *bhujā* and *koṭi* (areas of the squares whose sides are equal to *bhujā* and *koṭi*) is same as the (area of) square whose side is equal to the *karṇa*.

The proof of the *bhujā-koṭi-karṇa-nyāya* presented here is quite interesting in two respects — (a) it is an elegant dissectional proof that does not involve arithmetic, algebra or geometry, and (b) the reasoning offered by Nīlakaṇṭha in making a choice between two possible approaches for proving a general result of this kind. There is nothing much to labour here regarding the former as it would become apparent to the reader by going through the proof. However, as an introduction it is desirable to brief about the latter given, here Nīlakaṇṭha initially considers two different approaches — (i) *saṁkhyā* (enumerative), and (ii) *kṣetracchedana* (geometrical/dissectional) — for demonstrating the theorem. Then based on sound mathematical reasoning, he makes the choice of the latter to the former. Incidentally, the discussion provides a glimpse of the logical rigour with which the Kerala mathematicians approached science.

As mentioned earlier, both the theorem and its proof are quite old and can be traced back — though not in their present forms — to second millennium BCE, if not earlier. At this point the question that arises is — what exactly was the significance of supplying one more proof if there already existed several proofs? This question simply makes sense considering the fact that independent proofs are always welcome as they expand the horizon of thinking. Also, the joy of discovering a proof for oneself can be quite intense, even if one is not the first[21] in discovering a proof. Being first in proving a theorem, may have distinct advantages of getting name and fame. But nevertheless, being down the line and independently arriving at a proof, does not deprive the mathematician from experiencing the joy of mathematical discovery. That's precisely what could have provided the necessary impetus for Nīlakaṇṭha to come up with a proof of *bhujā-koṭi-karṇa-nyāya*.

STATEMENT OF THE PROBLEM

तत्र एकेनार्धेन अखिलक्षेत्रस्य गणितस्य बीजभूतं भुजाकोटिकर्ण-क्षेत्रगणितन्यायः प्रदर्श्यते –

दोः कोटिभुजयोर्योगः कर्णबाहुचतुर्भुजः। – इति।

यस्मिन् क्षेत्रे भुजाकोटिकर्णेषु त्रिषु, ज्ञाताभ्यां द्वाभ्यां, इतरेतरज्ञानोपायो जिज्ञास्यते, तत्र त्रीणि समचतुरश्रक्षेत्राणि[22] कल्प्यानि। तत्रैकस्य चत्वारो बाहवो जिज्ञासिताः क्षेत्रबाहुना तुल्याः, अन्यस्य कोट्या तुल्याः, अन्य(स्य)

---

21 Apparently as late as CE 1876, the twentieth American President, James Garfield has adduced an interesting geometrical proof of the Pythagorean Theorem using the properties of a trapezium.

22 The reading in the manuscript is *samacaturakṣetrāṇi*. It was felt that there could be error, and hence letter *śra* which is found missing has been introduced.

कर्णतुल्याः। तत्र कर्णतुल्यबाहुक्षेत्रम् इतरचतुरश्रद्वययोगसमम्। स्वसंवेद्य चैतत् प्रतिभाजुषाम्। कथम्?

tatra ekenārdhena akhilakṣetrasya gaṇitasya bījabhūtaṁ bhujākoṭikarṇakṣetragaṇitanyāyaḥ pradarśyate –
doḥ koṭibhujayoryogaḥ karṇabāhucaturbhujaḥ । – *iti*

yasmin kṣetre bhujākoṭikarṇeṣu triṣu, jñātābhyāṁ dvābhyāṁ, itaretarajñānopāyo jijñāsyate, tatra trīṇi samacaturaśrakṣetrāṇi kalpyāni । *tatraikasya catvāro bāhavo jijñāsitāḥ kṣetrabāhunā tulyāḥ, anyasya koṭyā tulyāḥ, anya(sya) karṇatulyāḥ* । *tatra karṇatulyabāhukṣetram itaracaturaśradvayayogasamam* । *svasaṁvedyaṁ caitat pratibhājuṣām* । *katham?*

Herein by a half [verse] it is demonstrated the rule relating the sides (*bhujā* and *koṭi*) of a right-angled triangle and its hypotenuse, which forms the basis for all the calculations based on geometrical constructions.

> The sum of the *bhujā* and *koṭi* (areas of the squares whose sides are equal to *bhujā* and *koṭi*) is the same as the (area of) square whose side is equal to the *karṇa*.

Consider the situation in which among the three sides of a triangle, *bhujā, koṭi* and *karṇa,* two are known and the third is desired to be known. Here, we need to construct three squares (*samacaturaśrakṣetra*). Of the three, one of them has all the four sides equal to the *bhujā,* for the other it is same as *koṭi* and for the third it is equal to *karṇa*. Now, the area of the square whose sides are equal to *karṇa,* is equal to the sum of the areas of the other squares. This is self-evident to the intelligent ones. How?

DEMONSTRATION OF THE RESULT IN THE CASE OF A SQUARE

तत्र समचतुरश्रक्षेत्रं वीक्षमाणस्य तत्पर्यन्तभागेभ्यः चतुर्भ्यस्तुल्येभ्यो भुजाख्येभ्यः एकान्तरितकोणयोः अन्तरस्य कर्णाख्यस्य दैर्घ्याधिक्यं प्रथमं स्फुरेत्। बाहुष्वपि मिथः संस्पृष्टैकाग्रयोर्द्वयोर्द्वयोः इतरेतरं विपरीतदिक्कृत्वे

च ज्ञाते कर्णद्वयस्य सामान्येन भाव्यम्, परिमाणभेदस्य कारणाभावात् - इति च स्फुरेत्।

तयोरेकतरकर्णदिगभिमुखस्य तस्याऽयामत्वाङ्गीकारे इतरकर्णाग्रद्वयस्य कटाक्षस्य तत्समायामविस्तारं कोणप्रतिकोणान्तसमबाहुकं पूर्वचतुरश्रगर्भं चतुरश्रान्तरं प्रतीयेत। तत्र यदि प्रथमस्य बाहवो दिगभिमुखाः, तर्हि द्वितीयस्य विदिगभिमुखाश्चत्वारः। तत्र प्रथमस्यैकैकं बाहुमभितो ये द्वे त्र्यश्रे उत्पद्येते, तत्र प्रथमचतुरश्रान्तर्गतमेकम्, तद्बहिर्गतमन्यत्। एवं प्रथमस्य प्रतिबाहु द्वे द्वे त्र्यश्रे स्तः। तान्यष्टौ सर्वाणि मिथस्तुल्यान्येव।

tatra samacaturaśrakṣetraṁ vīkṣamāṇasya tatparyanta-bhāgebhyaḥ caturbhyastulyebhyo bhujākhyebhyaḥ ekāntaritakoṇayoḥ antarasya karṇākhyasya dairghyādhikyaṁ prathamaṁ sphuret। bāhuṣvapi mithaḥ saṁspr̥ṣṭaikāgrayor-dvayordvayoḥ itaretaraṁ viparītadikkatve ca jñāte karṇadvayasya sāmānyena bhāvyam, parimāṇabhedasya kāraṇābhāvāt - iti ca sphuret।

tayorekatarakarṇadigabhimukhasya tasyā 'yāmatvāṅgīkāre itarakarṇāgradvayasya kaṭākṣasya tatsamāyāmavistāraṁ koṇapratikoṇāntasamabāhukaṁ pūrvacaturaśragarbhaṁ caturaśrāntaraṁ pratīyeta। tatra yadi prathamasya bāhavo digabhimukhāḥ, tarhi dvitīyasya vidigabhimukhāścatvāraḥ। tatra prathamasyaikaikaṁ bāhumabhito ye dve tryaśre utpadyete, tatra prathamacaturaśrāntargatamekam, tadbahirgatamanyat। evaṁ prathamasya pratibāhu dve dve tryaśre staḥ। tānyaṣṭau sarvāṇi mithastulyānyeva।

For any observer looking at the square, it is quite apparent that the length of the *karṇa*, which is equal to the distance of separation between any two alternative corners[23] of the square, is greater than the length of any of the four sides, which are equal in dimension. Morever, even among the four sides of the square which are connected as well as perpendicular to each other, it is quite apparent that the *karṇa*s should be of equal length, as there is no

[23] Choosing two corners, by leaving out a corner in between.

reason for them to differ in their dimension.[24]

Now, for a person looking along the direction of any one of the hypotenuses, the existence of another square can be visualized — circumscribing the first square — in such a way that the sides of this square, having a length equal to *karṇa* graces the tip of the *karṇa*. [In other words,] if the sides of the first square are parallel to the directions (east–west and north–south), then the four sides of the second square will be parallel to the intermediate directions (northeast-southwest and northwest-southeast). Now, (it can be seen that) two triangles are formed on either side of the four sides of the first square, of which one will be falling inside the first square and the other outside. Thus, there are two triangles associated with each side of the first square. All these eight triangles are similar to each other (see *fig.* 8.1).

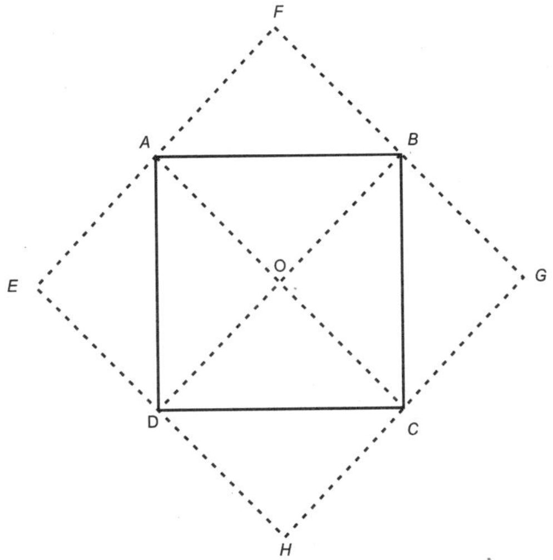

fig. 8.1: Geometrical demonstration of *bhujā-koṭi-karṇa-nyāya* when *bhujā* and *koṭi* are equal

24 The presence of perfect symmetry, or the absence of evidence on the contrary are taken to prove the point sometimes.

In *fig.* 8.1, *ABCD* is the square initially considered whose side is taken to be *a*. *EFGH* is the square which is visualized later whose side is equal to the *karṇa* of *ABCD*. It is obvious that the area of the bigger square will be equal to $2a^2$. Thus, the figure demonstrates the result that the area of any square — whose side is equal to the *karṇa* of the other — is twice the square of the side of the other.

तेषां प्रत्येकं फलं च प्रथमचतुरश्रफलचतुरंशतुल्यम्, कर्णाभ्यां चतुर्धा[25] विभक्तस्य प्रथमचतुरश्रस्य खण्डैश्चतुर्भिस्तुल्यत्वात्। तद्बहिर्गतानां च।[26] ततस्तयोरन्तर्गतचतुरश्रफलात् द्विगुणं बहिर्गतचतुरश्रफलम्। चतुरश्रक्षेत्रफलं च तद्बाहुवर्गसमम्। तस्मात् तत्तत्समचतुरश्रबाहुवर्गात् द्विगुणं तत्कर्णभुजक्षेत्रफलम्, फलमूलतुल्यं च बाहुदैर्घ्यम्। इत्येतत्सर्वं द्रव्यभूतस्य पुरुषस्य अवश्यं स्फुरति, तस्य संवेद्यसम्बन्धि-अखिलन्यायावयवानां तत्त्वज्ञानपर्यन्तं जिज्ञासाया अविश्रान्तेः।

teṣāṁ pratyekaṁ phalaṁ ca prathamacatura-śraphalacaturaṁśatulyam, karṇābhyāṁ caturdhā vibhaktasya prathamacaturaśrasya khaṇḍaiścaturbhistulyatvāt । tadbahirgatānāṁ ca । tatastayorantargatacaturaśraphalāt dviguṇaṁ bahirgatacaturaśraphalam । caturaśrakṣetraphalaṁ ca tadbāhuvargasamam । tasmāt tattatsamacatura-śrabāhuvargāt dviguṇaṁ tatkarṇabhujakṣetraphalam, phalamūlatulyaṁ ca bāhudairghyam । ityetatsarvaṁ dravyabhūtasya pusuṣasya avaśyaṁ sphurati, tasya saṁvedyasaṁbandhiakhilanyāyāvayavānāṁ tattvajñāna-paryantaṁ jijñāsāyā aviśrānteḥ ।

The area (*phalaṁ*) of each of these [triangles] (the ones inside the first square) is equal to one-fourth the area of the first square, because these triangles are none other than the four

25 The reading in the printed text is *caturthā*, which is obviously an error.

26 A sentence break here seems to be essential for proper understanding of the text, which is missing in the present edition.

(equal) parts of the square created by the intersection of the *karṇas*. [This is true] of the outer ones (triangles) also. Therefore, among those two, the area of the outer square is twice the area of the inner square. And the area of a square is equal to the square of its side. Therefore, the area of any square — whose side is equal to the *karṇa* of the other — is twice the square of the side of the other (whose *karṇa* is considered), and the length of its side is equal to the square root of the area. All these become evident to any serious (seriously thinking) person, since his urge to know (*jijñāsā*) never subsides till he clearly understands all the principles involved in what he wants to know.[27]

RELATION BETWEEN THE DIAGONAL AND THE SIDES OF A RECTANGLE

एवं समचतुरश्रकर्णज्ञानानन्तरमेव आयतचतुरश्रकर्णजिज्ञासा स्यात्। तस्य भुजाकोट्योः अतुल्यपरिमाणत्वात् तयोः कतरस्य वर्गं द्विगुणितं तत्फलं स्यादित्यत्र सन्देहच्छेदिविशेषहेत्वभावात् तद्योगार्धस्य द्विगुणीकरणे तदैक्यमेव तत्फलं स्यादिति स्फुरणस्य ऊहत्वात् तदेव वास्तवं फलम्, ततोल्पं वाधिकं वा, इति संशयः स्यात्। ततस्तन्निर्णयार्थं पुनरपि निरूपयति।

evaṁ samacaturaśrakarṇajñānānantarameva āyatacaturaśrakarṇajijñāsā syāt। tasya bhujākoṭyoḥ atulyaparimāṇatvāt tayoḥ katarasya vargaṁ dviguṇitaṁ tatphalaṁ syādityatra sandehacchediviśeṣahetvabhāvāt tadyogārdhasya dviguṇīkaraṇe tadaikyameva tatphalaṁ syāditi sphuraṇasya ūhatvāt tadeva vāstavaṁ phalam, tatolpaṁ vādhikaṁ vā, iti saṁśayaḥ syāt। tatastannirṇayārthaṁ punarapi nirūpayati।

It is only after understanding the (relation between the) diagonal (and the sides) of a square, does the desire to understand the same in the case of a rectangle (*āyatacaturaśra*) arise. Since in this case, the sides corresponding to the *bhujā* and *koṭi* are of unequal lengths,

27 This statement and the analysis that follows, not only mirrors the spirit of inquiry of Nīlakaṇṭha, but also his passion for science.

it is not possible to get rid of the doubt, as to whose square, when doubled will give the desired area.[28] Intuitively, one may say that the area will be twice the mean value of the squares of the sides; but then, since it is only a guess, again there is bound to be an uncertainty whether that will be the actual value, or it could be more or less than that. To establish that, it is further being explained.

TWO POSSIBLE WAYS OF REASONING

तत्र निरूपणस्य द्वौ मार्गौ। द्वयो राश्योर्वर्ग(योग)[29]मूलं निरवयवम् एकादिसंख्याविशेषेषु कयोः कयोः स्यादिति संख्या [निश्चीयते।][30] तदन्वेषण एको मार्गः। अशेषविषयव्यापी क्षेत्रच्छेदकद्वारः अन्यो मार्गः। प्रथममनेन मार्गेण गन्तव्यम् अन्यस्य (अर्थात् प्रथमस्य) अल्पविषयत्वात्, आनन्त्यात्, क्वचिदप्यविश्रान्तेश्च।

tatra nirūpaṇasya dvau mārgau। dvayo rāśyorvarga(yoga) mūlaṁ nirvayavam ekādisaṁkhyāviśeṣeṣu kayoḥ kayoḥ syāditi saṁkhyā [niścīyate।] tadanveṣaṇa eko mārgaḥ। aśeṣa-viṣayavyāpī kṣetracchedakadvāraḥ anyo mārgaḥ। prathamamanena mārgeṇa gantavyam anyasya (arthāt prathamasya) alpaviṣayatvāt, ānantyāt, kvacidapya-viśrānteśca।

There are two ways of demonstration. One is by counting and listing the set of natural numbers (*ekādisaṁkhyāviśeṣa*) which can be represented as the square root of the (sum of the) squares of two numbers. The other is to resort to the geometrical method which is all-encompassing (*aśeṣaviṣa-*

28 Area of the square whose side is equal to the diagonal of the rectangle.

29 The word *yoga* given in the parenthesis is found to be missing in the text. This is a serious omission, that could be due to scribe, editor, typo, and so on.

30 Here the material inside the square braces, seems to be an addition made by the editor, which is unwarranted. The text reads fine without that.

yavyāpī). First we choose the latter (geometrical approach), as in the case of the other (the former), the domain of applicability is limited, the set will be enormous in size and will be never-ending.[31]

DISSECTIONAL METHOD

तत्र कथं छित्वा योजने द्वित्रा वा सर्वे वा बाहवो जिज्ञासितकर्णतुल्याः स्युः? तत्र कोटितुल्यचतुरश्रस्य एकेन बाहुना अल्पस्य बाहुं कुतश्चित्कोणतः प्रभृति सन्धाय संश्लिष्टस्य क्षेत्रस्य यत्र कोणयोः संयोगः स भागो दोःकोटितुल्यः। महति चतुरश्रे संश्लिष्टाद् बहिर्गतः खण्डो दोःकोट्यन्तरसमः। इतरौ बाहू कोटितुल्यावेव। अल्पस्य चान्येन संस्पृष्टौ बाहू दोःसमावेव। एवं च पर्यन्तभागाः षोढा विभक्ताः स्युः।

tatra kathaṁ chitvā yojane dvitrā vā serve vā bāhavo jijñāsita-karṇatulyāḥ syuḥ? tatra koṭitulyacaturaśrasya ekena bāhunā alpasya bāhuṁ kutaścitkoṇataḥ prabhr̥ti sandhāya saṁśliṣṭasya kṣetrasya yatra koṇayoḥ saṁyogaḥ sa bhāgo doḥkoṭitulyaḥ । mahati caturaśre saṁśliṣṭād bahirgataḥ khaṇḍo doḥkoṭyantarasamaḥ । itarau bāhū koṭitulyāveva । alpasya cānyena saṁspr̥ṣtau bāhū doḥsamāveva । evaṁ ca paryanta-bhāgāḥ ṣoḍhā vibhaktāḥ syuḥ ।

[If so] by splitting and combining in which way, shall we be able to show that one or two or all the sides (of the resulting geometric figure) will be equal to the desired hypotenuse? Now, place the side of the smaller square adjacent to one of the sides of the *koṭi* square (larger one), in such a way that corners coincide. In the resulting composite figure, the length

31 What is noteworthy is the choice of Nīlakaṇṭha, of the geometrical method to the enumerative one. Notwithstanding the fact that the number of observations of the particular instances, could be very large (infinite !), it is not going to be complete (ending). The word *viśrānti* means taking rest. In this context, it refers to resting after listing sets of three numbers ("Pythagorean" triplets). That the list of elements (set of three numbers), in this set is not finite, is indicated by the use of the word *aviśrānti* (never-ending).

of the side where the corners coincide is equal to the sum of the *doḥ* and *koṭi*. In the larger square, the length of the side which lies outside the region of overlap, is equal to the difference between the *doḥ* and *koṭi*. The other two sides (of the larger square) is same as *koṭi*. The two sides of the smaller one, which are in contact with the other (larger) is equal to *doḥ*. This is the sixfold distribution of the sides of the resulting composite figure (see *figs*. 8.2(a) and (b)).

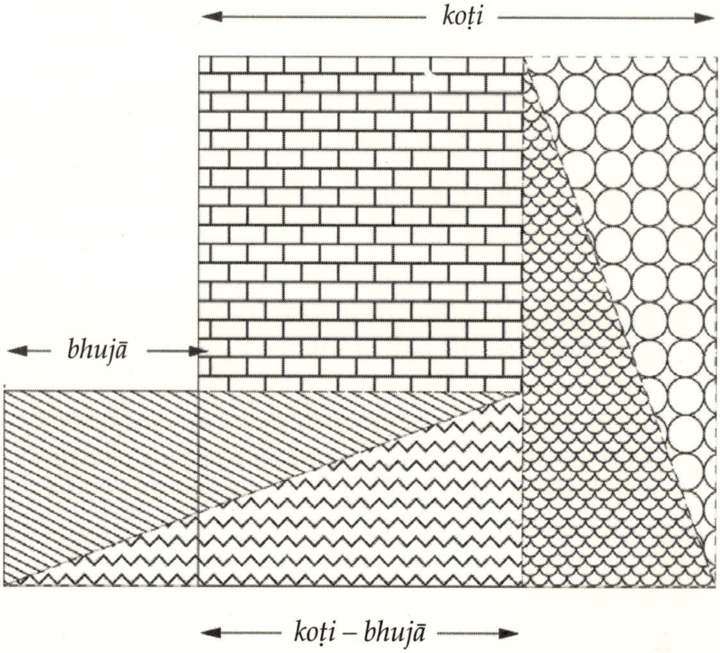

fig. 8.2(a): Dissectional method to prove *bhujā-koṭi-karṇa-nyāya*

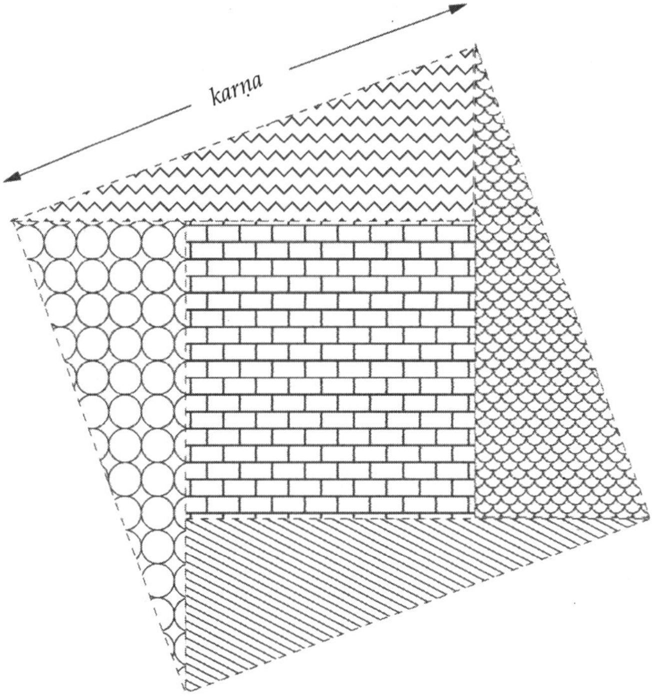

fig. 8.2(b): Dissectional method to prove *bhujā-koṭi-karṇa-nyāya*

In *fig*. 8.2(a), two squares of sides *bhujā* and *koṭi* are placed side by side. The interface between different patterns represents the directions along which the figure is to be cut. *Fig*. 8.2(b) shows the re-assembled form of *fig*. 8.2(a), by sliding the triangles. They are placed so as their hypotenuses face outside, demonstrating the desired result — "The sum of the *bhujā* and *koṭi* (areas of the squares whose sides are equal to *bhujā* and *koṭi*) is the same as the (area of) square whose side is equal to the *karṇa*."

तत्र कोणयुतेर्महति दोःकोट्यन्तरसमे लाञ्छने कृते दोःकोटियोगतुल्यस्य बाहोर्लाञ्छनाद् बहिर्गतो भागो दोःसमः, अन्यः कोटिसमः। महतोऽन्तरे-

णोनत्वात्, अल्पस्यान्तरेणाधिकत्यात् कोटितुल्यत्वञ्च तयोः। महच्चतुरश्रं लाञ्छनमार्गेण छिन्द्यात्। तदा तदायतचतुरश्रं दोःसमविस्तारं कोटिसमायामं च, इतरस्य खण्डस्याश्रे द्वे कोटिकोटितुल्ये, द्वे चतुरश्रतुल्ये, अन्ये दोःकोट्यन्तरतुल्ये।

तस्य शिखरे छित्त्वा पृथक्कृतः खण्डः दोःकोट्यन्तरसमचतुरश्रं क्षेत्रम्। तच्छिष्टं चायतचतुरश्रं पूर्वेण तुल्यम्। तयोर्द्वयोः कर्णमार्गेण छिन्नयोः चत्वारि त्र्यश्राणि स्युः। तत्कर्णाश्चत्वारो जिज्ञासितकर्णतुल्या एव। तानि तथा सन्दध्यात् यथा तत्कर्णा बहिर्गताः स्युः। तथा तन्मध्यत्र्यश्रशून्यं यद्विवरं तद् दोःकोट्यन्तरसमचतुरश्रम्। ततस्तत् तृतीयेन खण्डेन पूरणीयम्। तथा तत् जिज्ञासितकर्णसमबाहुकं समकर्णसमचतुरश्रं स्यात्। अतः सिद्धम् – दोःकोटिभुजयोर्योगः कर्णबाहुचतुर्भुजः – इति।

tatra koṇayutermahati doḥkoṭyantarasame lāñchane kṛte doḥkoṭiyogatulyasya bāhorlāñchanād bahirgato bhāgo doḥsamaḥ, anyaḥ koṭisamaḥ। mahato 'ntareṇonatvāt, alpasyāntareṇādhikyāt koṭitulyatvañca tayoḥ। mahaccaturaśraṁ lāñchanamārgeṇa chindyāt। tadā tadāyatacaturaśraṁ doḥsamavistāraṁ koṭisamāyāmaṁ ca, itarasya khaṇḍasyāśre dve koṭikoṭitulye, dve caturaśratulye, anye doḥkoṭyantaratulye।

tasya śikhare chittvā pṛthakkṛtaḥ khaṇḍaḥ doḥkoṭyantarasamacaturaśraṁ kṣetram। tacchiṣṭaṁ cāyatacaturaśraṁ pūrveṇa tulyam। tayordvayoḥ karṇamārgeṇa chinnayoḥ catvāri tryaśrāṇi syuḥ। tatkarṇāścatvāro jijñāsitakarṇatulyā eva। tāni tathā sandadhyāt yathā tatkarṇā bahirgatāḥ syuḥ। tathā tanmadhyatryaśraśūnyaṁ yadvivaraṁ tad doḥkoṭyantarasamacaturaśram। tatastat tṛtīyena khaṇḍena pūraṇīyam। tathā tat jijñāsitakarṇasamabāhukaṁ samakarṇasamacaturaśraṁ syāt। ataḥ siddham — doḥkoṭibhujayoryogaḥ karṇabāhucaturbhujaḥ — iti।

In the larger square, if we mark a point, from the point of intersection of the edges, at a distance equal to the difference in *bhujā* and *koṭi*, the length of portion — on that side which is equal to the sum of the *doḥ* and *koṭi* — which lies outside the mark is equal to *doḥ* and the other is equal to *koṭi*. In fact, both can be seen to be equal to *koṭi* as in one case the larger one is diminished by the difference and the smaller one is added to the difference. Let the larger square be cut along the direction of the mark. Now, the resulting rectangle will have length and breadth equal to *doḥ* and *koṭi*. In the figure that remains, two sides are the equal to *koṭi* in length, two are equal to the square (*doḥ*-square), and the other two have lengths equal to the difference between *doḥ* and *koṭi*.

If we chop off the top portion of this figure, then the portion that is obtained will be a square whose sides are equal to the difference between *doḥ* and *koṭi*. The remaining portion of the figure will be identical with the rectangle obtained earlier. If these two (rectangular figures) are cut along the diagonals, we obtain four triangles. The *karṇa*s of these triangles are same as the *karṇa*s that were desired to be found out. Now, place them (the four triangles) such that their *karṇa*s lie outside. Thus (after placing them), the portion/gap at the centre, which is not overlapped by the triangles, happens to be a square with sides equal to the difference between the *doḥ* and *koṭi*. This gap should be filled in by the third piece. Thus, we have obtained a square whose sides are equal to the *karṇa* desired to be known. Hence, it has been established that — "The sum of the *bhujā* and *koṭi* (areas of the squares whose sides are equal to *bhujā* and *koṭi*) is the same as the (area of) square whose side is equal to the *karṇa*" (see *fig*. 8.2(b)).

CONCLUDING REMARKS

दोःसमचतुर्बाहुकस्य कोटिसमचतुर्बाहुकस्य च चतुरश्रयोरेकीकरणे चतुर्भुजं कर्णबाहुकं स्यात्। इत्येतत्सर्वं युक्तिमूलमेव, न त्वागममूलम्। ततः प्रतिभाजुषां चक्षुर्हस्तादिव्यापारं विना मनसा निरूप्य एवं निर्णेतुं शक्यम्। अनयैव दिशा विष्कम्भपरिधिज्याचापसम्बन्धश्च निरूप्यैव ज्ञेयः। एतत्सर्वं मया आर्यभटीयव्याख्याने प्रपञ्चितमिति विरम्यते।

doḥsamacaturbāhukasya koṭisamacaturbāhukasya ca caturaśrayorekīkaraṇe caturbhujaṁ karṇabāhukaṁ syāt।

> *ityetatsarvaṁ yuktimūlameva, na tvāgamamūlam* | *tataḥ pratibhājuṣāṁ cakṣurhastādivyāpāraṁ vinā manasā nirūpya evaṁ nirṇetuṁ śakyam* | *anayaiva diśā viṣkambha-paridhijyācāpasaṁbandhaśca nirūpyaiva jñeyaḥ* | *etatsarvaṁ mayā āryabhaṭīyavyākhyāne prapañcitamiti viramyate* |

> Thus if the squares of sides equal to the *doḥ* and *koṭi* are combined to form one composite figure [appropriately], then the resultant figure is a square with sides equal to the *karṇa*. All this is based on reasoning alone and does not need the authority of any authentic text [for its validity]. Hence, the intelligent ones will be able to convince themselves (with this relation), mentally without even employing their eyes, hands, etc. (for drawing figures). The relation between the diameter and the circumference, and that of the chord and the arc can be established along the same lines. Since all this has been explained by me in *Āryabhaṭīya-Bhāṣya* in great detail, I stop here at this point.

Among other things in the above passage we would particularly like to highlight two statements through which Nīlakaṇṭha conveys very important messages to budding scientists and researchers, irrespective of their disciplines of study.

> एतत्सर्वं युक्तिमूलमेव, न त्वागममूलम्।
>
> *etatsarvaṁ yuktimūlameva, na tvāgamamūlam* |

> All this is based on reasoning alone and does not need the authority of any authentic text [for its validity].

Drilling on text-based material may develop formal ability in a student but does not bring out inherent capacity to the maximum possible extent. It is perhaps with this in mind, Nīlakaṇṭha strongly recommends independent thinking.[32]

---

[32] Which had played the key role in almost all the major discoveries of the past.

Another important implication of the message seems to be —

> It would be interesting to know what is true; and more exciting would be to know why it is true; but it will be all the more thrilling to know why it is true by discovering the truth for oneself.

> ततः प्रतिभाजुषां चक्षुर्हस्तादिव्यापारं विना मनसा निरूप्य एव निर्णेतुं शक्यम्।
>
> *tataḥ pratibhājuṣāṁ cakṣurhastādivyāpāraṁ vinā manasā nirūpya eva nirṇetuṁ śakyam* ।

> Hence, the intelligent ones will be able to convince themselves (with this relation) mentally, without even employing their eyes, hands, etc.

There seems to be the suggestion in the above statement to try and work out things mentally — to the extent possible — rather than trying to resort to the external means in the very first instance. The word, etc. after eyes and hands take into account all other tools such as pen, paper, pencil and not to leave calculators and computers employed these days even to do the elementary calculations.

References

SOURCE WORKS

{*ABB*} *Āryabhaṭīya* of Āryabhaṭācārya with the *Mahābhāṣya* of Nīlakaṇṭha Somasutvan, Part II *Kālakriyāpāda*, ed. Sāmbaśiva Śāstrī, Trivandrum Sanskrit Series 110, Trivandrum, 1931.

{*GVV*} *Grahasphuṭānayane Vikṣepavāsanā* of Nīlakaṇṭha Somayājī, ed. in *Gaṇitayuktayaḥ* by K.V. Sarma, Vishveshvaranand Vedic Research Institute, Hoshiarpur, 1979.

{*GYB*} *Gaṇita-yukti-bhāṣā* of Jyeṣṭhadeva, ed. and tr. K.V. Sarma with explanatory notes by K. Ramasubramanian, M.D. Srinivas and M.S. Sriram, New Delhi: Hindustan Book Agency, 2008; repr. Springer, 2009.

[*ISD*] *Siddhānta-Darpaṇa* of Nīlakaṇṭha Somayājī with auto-commentary, ed. K.V. Sarma, Panjab University, 1977.

{*TS*} *Tantrasaṁgraha* of Nīlakaṇṭha Somayājī tr. with mathematical notes by K. Ramasubramanian, M.S. Sriram, jointly published by Delhi: Hindustan Book Agency, and London: Springer, 2011.

SECONDARY WORKS

Duke, Dennis, 2005, "The Earth in India: The Mathematical Basis of Ancient Indian Planetary Models", *Arch. Hist. Exact. Sc.*, **59:** 563-76.

Narasimha, Roddam, 2007, "Epistemology and Language in Indian Astronomy and Mathematics", *J Indian Philos*, **35:** 521-41.

Ramasubramanian, K., M.D. Srinivas, 2010, "Development of Calculus in India", in *Studies in the History of Indian Mathematics*, ed. C.S. Seshadri, New Delhi: Hindustan Book Agency, pp. 201-86.

Ramasubramanian, K., M.D. Srinivas and M.S. Sriram, 1994, "Modifications of the Ancient Planetary Model by the Astronomers of the Kerala school and the Implied Heliocentric Picture of the Planetary Motion", *Current Science*, **66:** 784-90.

Sarma, K.V., 1972a, *A Bibliography of Kerala and Kerala-based Astronomy and Astrology*, Hoshiarpur: Vishveshvaranand Vedic Research Institute.

———, 1972b, *A History of the Kerala School of Hindu Astronomy*, Hoshiarpur: Vishveshvaranand Vedic Research Institute.

Srinivas, M.D., 2005, "Proof in Indian Mathematics", G. Emch, M.D. Srinivas and R. Sridharan (eds.), *Contribution to the History of Mathematics in India*, New Delhi: Hindustan Book Agency, 209-48.

Sriram, M.S., K. Ramasubramanian and M.D. Srinivas (eds.), 2002, *500 Years of Tantrasangraha: A Landmark in the History of Astronomy*, Shimla: Indian Institute of Advanced Study.

Contributors

Chidambaram, Raju, Ph.D., had his early education in Kerala, later earning his BA (Hons.) in Mathematics from the University of Madras and MS in Statistics from the Indian Statistical Institute, Calcutta. He received his Ph.D. in Operations Research from the Case Western Reserve University, Cleveland, Ohio. A large part of his professional career was working in satellite communications at COMSAT and INTELSAT in Washington DC. After his wife Shobha introduced him to Chinmaya Mission, he took a deep interest in learning Advaita Vedānta from Swami Chinmayananda and other *ācāryas* of the Chinmaya Mission. Retiring in 1995, Raju became curious about the possibility of applying mathematical methods to the science of spirituality taught in our Upaniṣads. The outcome was a series of papers published from 2000 through 2008, culminating in the book *Law of Love and the Mathematics of Spirituality* (Author House, Indiana, in 2011) [*address*: 1406 Greenwood Place, ALEXANDRIA, VA 22304-1604 (USA); *E-mail*: <aiyers@comcast.net>].

Chidambaram, Shobha, MD, received her medical education in Nagpur, India and later completed residency programme in Neurology at the Howard University Hospital, Washington DC. She specializes in adult neurology serving inpatients in several area hospitals and outpatients from her office in Oxon Hill, Maryland. Her abiding interest in Vedānta was sparked during her college days in Nagpur after listening to talks by Swami Chinmayananda [*address*: 1406 Greenwood Place, ALEXANDRIA, VA 22304-1604 (USA); *E-mail*: <shobhac@comcast.net>].

Deshmukh, Vinod D., MD, Ph.D., is a retired neurologist from Jacksonville, Florida. He was an Associate Professor of Neurology at University of Florida. He also had his own private practice at the Flagler Hospital, in St Augustine, Florida. He has published over sixty scientific articles including a monograph on Brain Blood Flow.

His recent articles were on the Neuroscience of Meditation, the Multistream Self, Turīya, Prāṇa-Dhyāna, Vedic Psychology as a Science of Wisdom, and The Cognitive-Pause and Unload, the CPU Hypothesis of Meditation and Creativity. His latest book is called *The Astonishing Brain and Holistic Consciousness: Neuroscience and Vedanta Perspectives* (2012). Poetry, photography and oil painting are his hobbies. He has published four books of his original poetry, essays and photographs [*address*: 3600 Rustic Lane Jacksonville, FLORIDA 32217 (USA); *E-mail:* <vinod38@aol.com>].

Jatavallabhula, Ravi, is an author, researcher, and works as a project manager at a reputed US company. He holds a masters degree from Indian Institute of Science, Bangalore. Ravi has published books, articles and presented at various conferences in the area of Sanātana Dharma and stress-free living. Having born in a Vedic family and being a grandson of famous Sanskrit scholar late Mr Purushottam Jatavallabhula, Ravi explores into Vedic knowledge and interprets the scriptures using modern science concepts [*address*: 2549 Cobble Hill Ct, Unit F, SAINT PAUL, MN 55125 (USA) or 36-3-4 Marla Lane, Temple Street, KAKINADA - 533 001 (AP); *E-mail*: <rjata@genuges.com>, <jataravi@yahoo.com>].

Kak, Subhash, is Regents Professor of Computer Science at Oklahoma State University in Stillwater. His research includes archaeo-astronomy and history of science and art. He has made the surprising discovery that the ancient Indians knew that the sun and the moon are approximately 108 times their respective diameters from the earth. This knowledge was coded into temple architecture, in the 108 poses of Indian classical dance, and the 108 prayer beads of the *japamālā* [*address*: 2605 W. Summerlin Ct., STILLWATER, OK 74074 (USA); *E-mail*: <subhashkak@gmail.com>].

Mahesh, K., a scholar of Siddhānta Jyotiṣa, has been doing research in the area of Ancient Indian Astronomy and Mathematics. His research papers were published in refereed journals as well as in international seminar proceedings. He got Ph.D. from IIT Bombay under the guidance of Prof. Ramasubramanian. Also he did his post-doctoral research at CNRS, Paris. Presently he is working for Sanskrit Promotion Foundation, New Delhi [*address*: Sanskrit Promotion Foundation, 11204/05, Gaushala Marg, Karol Bagh, NEW DELHI - 110 005; *E-mail*: <k.mahesh.iit@gmail.com>].

Prasad, M.G., is professor of mechanical engineering and the director of the noise and vibration control laboratory at Stevens Institute of Technology, Hoboken, New Jersey, USA. He has published more than 100 articles in journals and conference proceedings in the areas of acoustics, vibration and noise control. He has also presented several papers on Hinduism and acoustics at several conferences both in USA and India. Dr Prasad is a Fellow of the American Society of Mechanical Engineers, a Fellow of the Acoustical Society of America and a Life-Fellow of the Acoustical Society of India. In November 2013, Dr Prasad was invited to deliver the M.S. Narayanan memorial lecture on "Acoustics of Chants, Conch-Shells, Bells and Gongs in Hindu Worship Spaces" at ACOUSTICS 2013, an international conference in acoustics jointly organized by the Acoustical Society of India and the French Acoustical Society in New Delhi, India [*address*: Department of Mechanical Engineering, Stevens Institute of Technology, Hoboken, NEW JERSEY 07040, (USA); *E-mail*: <mprasad@stevens.edu>].

Ramasubramanian, K., holds a Ph.D. in theoretical physics, besides a Bachelors in engineering and a Masters in Sanskrit. His areas of interest include non-linear dynamics, Indian astronomy and mathematics as well as Indian philosophy. For completing a rigorous course in Advaita Vedānta he was honoured with the coveted title "Vidvat Pravara" in 2003 by the Ācārya of Śṛṅgerī. He was also conferred in 2008 the prestigious Maharshi Badarayan Vyas Samman by the President of India, in recognition of his outstanding contribution towards synergy between tradition and modernity [*address*: Department of HSS, IIT Bombay, Powai, MUMBAI - 400 076; *E-mail*: <mullaikramas@gmail.com>].

Sengupta, Sukalyan, is Professor & Chairperson, Civil & Environmental Engineering Department, and Director of the Center for Indic Studies at the University of Massachusetts Dartmouth. Besides his research expertise in environmental engineering — primarily global water sustainability and water/wastewater treatment and resource recovery — he is interested in consciousness studies, especially from a Vedāntic perspective [*address*: Professor & Chairperson, Civil & Environmental Engineering Department, Director, Center for Indic Studies University of Massachusetts Dartmouth, 285 Old Westport Road, NORTH DARTMOUTH, MA 02747 (USA); *E-mail*: <ssengupta@umassd.edu>].

Singh, Bal Ram, is Executive Mentor at School of Indic Studies at UMass Dartmouth, and Director, Botulinum Research Center, Institute of Advanced Sciences, Dartmouth. Dr Singh is a Professor of Biophysical Chemistry and Henry Dreyfus Teacher-Scholar at UMass Dartmouth and has been conducting research for twenty years on the molecular mode of action of botulinum and tetanus neurotoxins, and lately also on yoga, mind and consciousness. His research on biodefence and biotechnology is internationally recognized, and he has served on many national and international scientific panels. Dr Singh has published about 200 research articles; three dozen scholarly articles on issues related to Indian tradition, culture and philosophy; has edited six books, including *India's Intellectual Traditions and Contributions to the World* (2010); and has obtained nine patents. Dr Singh has been honoured with awards by several reputed institutions in India and abroad [*address*: Executive Mentor, School of Indic Studies, Professor and Director, Botulinum Research Center, Institute of Advanced Sciences, 78-540 Faunce Corner Mall Road, DARTMOUTH, MA 02747 (USA); *E-mail*: <bsingh@umassd.edu>].

Singh, Rishi Bal, is a Ph.D. student of Mechanical Engineering at the University of Illinois, Urbana-Champaign. He completed his Bachelor of Science degree in Mechanical Engineering at the University of Massachusetts Amherst in 2012. Singh has keen interest in Indian philosophy and has studied Sanskrit as an undergraduate student [*address*: Mechanical Science and Engineering, University of Illinois at Urbana-Champaign, Mechanical Engineering Building, 1206 W. Green St. URBANA, IL 61801 (USA); *E-mail*: <rishi.b.singh@gmail.com>].

Solfvin, Jerry, Ph.D., is Adjunct Associate Professor with Center for Indic Studies at University of Massachusetts Dartmouth. Dr Solfvin has practised, taught and researched the therapeutic and transcendent dimensions of meditation for over three decades. He is a full member, since 1975, of the Parapsychological Association. He is an award-winning teacher, a research consultant for individuals and groups, and an investigator with special interests in complementary and alternative medicines [*address*: 14 Florence Street, WESTPORT, MA 02790 (USA); *E-mail*: <jsolfvin@umassd.edu>].

Index

Absolute Being-in-Reality, 77
absorption, subjective experience of, 119
abstract concept of sameness, 53
abstract conceptualization, 50
achievement of consciousness normalization, 224
acoustical
  counterparts, 277
  environment, 280
acoustical aspects of
  language, 276
  literature, 276
  music, 269
  Sanskrit, 276
acoustics
  field of, 279
  in philosophy, 277
  in spirituality, 277
activation, spatiotemporal patterns of, 56
adaptive conscious behaviour, evolution of, 109
adaptive organization
  in neural systems, 58
  question of, 57
adaptive reality–consciousness, 124
Ādi Śaṅkarācārya Ātmabodha, 128
Advaita, 33
  Vedānta, 75
advanced theoretical physics, 175
advancement of human civilizations, 173
āgamamūla, 295
agency of space and time, 238
age-old introspective observations, 108
agnosticism, 19
Agrawal, M.M., 4, 243
aham brahmāsmi, 242
ahargaṇa, 299
āhata nāda, 263
ahaṁkāra, 72, 84-85, 87
Aitareya Upaniṣad, 242
Aiyar, K.N., 262
ākāśa, 239, 277
ākāśic records, concept of, 139
akhaṇḍa, 124
all-pervading cosmic energy, 240
alternative medicine, effectiveness of, 166
altruistic way of living, 70
America's
  collective mythology, 18
  radical dualism, 20
American Psychological

Association, 70
Amṛtabindu Upaniṣad, 262
amṛtānubhava, 69
anāhata nāda, 113, 263
ānanda, 88
ānandita, 123
Ananthu, T.S., 6-7, 245
anatomy, subtle differences in, 58
ancient
- Eastern philosophies, 131
- home of Aryan race, 259
- Indian texts, 10, 66
- philosophy of life, 69

ancient knowledge,
- body of, 258
- intricacies of, 5

animal
- behaviour, 53
- intelligence, 27, 29, 51, 53
  - experiments of, 52

antaḥkaraṇa, 72
antara, 83
antarātmā, 83
antarendriya(s), 83-84
antaryāmī, 77
aparokṣānubhūti, 69, 76
apauruṣeya, 74
apavarga, 86
arithmetic paradox, 47
Arjuna, 206
artificial intelligence (AI), 43, 46, 48, 53, 66
Āryabhaṭa, 300
Āryabhaṭīya, 287, 298
Āryabhaṭīya-Bhāṣya, 283-84, 286, 291, 297, 314
Aryan race, ancient home of, 259
asamprajñāta, 88
aśeṣaviṣayavyāpī, 308
asmitā, 88
assimilation of experiences, 101
associative priming, conditions of, 63
assumption of separateness, 38
astronomical procedures, 290
astronomy, history of, 283
Atharvaveda, 74, 105, 261
ātmā, 207
ātma-bhāva, 121
ātma-bodha, 69, 93
ātmajñāna, 83, 113, 144
ātmakatvam, 113
ātman, 6, 73-75, 78, 93-94, 113, 116, 153
- spiritual level of, 269

ātma-prasāda, 74, 112, 121
ātmasākṣātkāra, 87
ātmatuṣṭa, 112
ātmonnati, 120
autobiographical
- memory, 102
- self, 102

automatic homeostatic mechanisms, 109
autonomic nervous system (ANS), 189, 222
autonomy, old paradigms of, 47
autopoietic system, 98
avidyā, 153
aviśiṣṭa-mandakarṇa, 298
aviśrānti, 295
awareness
- of consciousness, 122, 248
- frame of, 236

ayana-calana, 291, 298, 300

āyatacaturaśra, 307
Āyurveda, 9

Bahadur, K.P., 277
bāhyātmā, 83
Barlowe, David H., 28
Barton, A., 8, 247
basic
    definition of consciousness, 233
    ingredients of mind, 107
    oneness nature of continuum, 194
    teachings of Vedānta, 137
    wakeful awareness, re-establishment of, 99
basis of Yoga, 86, 115
beam of intelligence, 154
behaviour
    of consciousness, 219
    experimental analysis of, 28
    temporal organization of, 103
behaviourism, 36
Being, realization of, 74
being-in-existence, modes of, 107
Bentov, I., 237
Bernard, Claude, 108, 110
Bhagavad-Gītā, 69, 110, 112, 115-16, 121, 160, 200, 205, 207
Bhāgavata Purāṇa, 267, 272
Bharati, Swami Paramananda (T.S. Shankara), 278
Bhāskara, 300
Bhaṭadīpikā, 298
bhujā, 303, 307, 311-12
bhujā-koṭi-karṇa-nyāya, 291, 294-95, 299-302, 305, 310-11
bhūtasaṁkhyā system, 296
Big Bang Singularity, 238
bījākṣaras, 272
biographies of many great saints, 165
biological intelligence, 49, 66
blindsight patients, 65
bliss, source of, 262
blissful
    conscious, integrative mode of, 68
    self-existence, sense of, 73
Block Universe, 140-41
body of ancient knowledge, 258
Bohm, 241
Brahmā (Lord), 251, 258
    is consciousness, 242
Brahmagupta, 300
Brahman, 6, 43, 94, 117, 125, 131, 138, 152, 240, 245, 267
    manifestation of, 242
    realization of, 73
Brāhmaṇas, 42
Brahma-Sūtra, 69
Brahmavarchas, 279
Brāhmībhāva, 73
brāhmī-sthiti, 70, 74, 121
brain
    function, quantum theoretic models of, 54
    scientific study of, 88
brain's self-organization, 79
brain-machine, 44
brainstem–cortical arousal system, 107
brainstem–limbic system, 88
brainstem–reticulo–thalamic arousal system, 107
Bṛhadāraṇyaka Upaniṣad, 77, 242
Broca's area, 89

Brown, G. Spencer, 35
Buddha, 123
Buddhism, 68-70
Buddhist
  sources, 18
  way of mindfulness meditation, 69

caitanya, 83
cakras, 272
Callender, Craig, 131
Candracchāyā-gaṇita, 286
Cannon, Walter B., 108, 110
Capra, Fritjof, 131
Cartesian co-ordinates, 143
caste morphogenesis, 59
Cayce, Edgar, 139
ceaseless wonderment, sense of, 68
Center for Inquiry, 27
central nervous system (CNS), 222
central thalamic nuclei, 82
centralization of prediction, 103
centre of
  consciousness normalization, 189
  motion of planet, 291
centrencephalic regions, 99
Cha, K.Y., 31
Chamu, S.V., 279
Chanana, Raj D., 175, 192
Chandas, 260
characteristic feature of consciousness, 231
Chidambaram, Raju, 9, 138, 145, 147, 167
Chinmayananda (Swami), 132, 134, 137, 144, 156, 164
Christian terminology, 6, 245
Church
  authorities, 5
  dark shadows of, 3
citta, 72, 84-85, 87
cittavṛtti, 72
classical
  computing, 47
  Hindu text, 266
  Indian literature, 276
  literature of Hinduism, 257
  physics, 132
  view, 133
    of universe, 132
Claus, Peter J., 19
climatic reasons, 2
clinical neurology, 97
cloak of darkness, corner of, 300
coccygeal triangle, 268
cognition, dominant mode of, 18
cognitive
  ability, 66
  capabilities, 53
  excellence, 71
  modality, 82
  scientists, 45
  style of all animals, 53
collection of *mantras*, 257
commissurotomy, 64
common belief, dualistic models of, 47
common sense notions of time and space, 141
communication
  mechanisms, 60
  process, 60

compartmentalization of subfields of science, 13
complete consciousness, state of, 232
complex neural processes, regulation of, 56
complex topic of consciousness, 74
computation of
  planetary longitudes, 287
  planetary positions, 288, 298
computational models, 49
computations of planetary longitudes, latitudes, eclipses, 288
concept(s) of
  ākāśic records, 139
  bio-psycho-social-existential self and mind, 104
  dharma traditions, 9
  fraction-living continuum, 187
  human consciousness, 80
  māyā, 229
  presence, 96
  subconscious mind, 86
  Uncle Sam, 104
conceptualization of mind, 48
conclusion of modern science, 135
conditions of associative priming, 63
conflicts between science and spirituality, 227
conscious
  arousal, 81, 92
  attentiveness, 124
  awareness, 82, 92, 97
    of self, 111
  experience, 65, 98
  sensations, 65
consciousness, 147
  awareness of, 122, 248
  basic definition of, 233
  behaviour of, 219
  characteristic feature of, 231
  coefficient, 232, 236, 254
  complex topic of, 74
  constant, 255
  cosmic levels of, 234
  definition of, 80
  evolutionary state of, 230
  fourth state of, 74
  fundamental questions of, 232
  higher level of, 252
  host, nature of, 209
  in Vedānta, overarching role of, 131
  infinite dimensions of, 248
  influence of, 66
  is all-pervasive, 248
  is *Brahman*, 156, 242
  merges with the Ultimate, 238
  model of, 248
  nature of, 153
  new model of, 94
  normalization, 208, 216, 221, 224, 228
    achievement of, 224
    centre of, 189
    purpose of, 212
  objects of, 234
  phenomenon of, 42
  question of, 42, 46
  realization, 215
  reunion of, 226
  scientific analysis of, 231
  study of, 54
  theory of, 43, 45, 66
  three states of, 191
  universal field of, 66
continual correction of dualis-

tic conceptions, 33
continuous perceptual–cognitive flow, 101
continuum,
  basic oneness nature of, 194
  cyclic nature of, 183, 206
  nature of, 216
  oneness of, 193
  particle nature of, 187
  singular nature of, 229
  state of, 198
  virtual nature of, 208, 220
contribution of Vedic knowledge, 8, 247
core consciousness, 102
core self, 102
corner of cloak of darkness, 300
correct geometric model for time, 249
correspondence between the Vedāntic concepts and Relativity Theory, 142
cortical structure, 58
cosmic
  interval, 150
  levels of consciousness, 234
  mind, 153, 158
    Vedāntic concept of, 134
  time, 148-51
cosmological
  model, 284
  singularity, 240
  theory of Vedānta, 130
cosmos, sentient *jīva*s of, 136
counterintuitive illusions, 48
Court of Appeals, 31
Coxeter, H.S.M., 176, 178
Crick, Francis, 7, 235, 246-47
cultural
  and philosophical differences, 3
  influence, 2
  traditions, 2
culture of
  Eastern world, 2
  Western world, 2
current
  consciousness, 81
  science and technology, 177
cycle of *yuga*s, 251
cyclic
  behaviour, 186
  nature of continuum, 183, 206

Dalela, A., 279
Damasio, Antonio, 102, 109
Dāmodara, 283, 285
  grace of, 285
Darwin, Charles, 27
data collection methodologies, 13
Davidson, R.J., 236
de Broglie, Louise, 141
deep sleep state, 192, 224
default mode neural network, 124
definition of
  consciousness, 80
  modern science, 244
  Paramātmā, 200
degree of spiritual detachment, 138
Democrat(s), 13, 19
Democratic, 20
depths of mystery, 38
description of motion of equinoxes, 298
desire for peace and tranquillity, 174

determinative faculty, 154
determinism, old paradigms of, 47
dharma, 3, 175, 193, 207
  interpretation, 198
    book, 200
  paradigm, 8
  scriptures, 175
  traditions, concept(s) of, 9
  major scriptures of, 193
  tradition of, 9
dhyāna,
  practice of, 114
  yoga practices of, 10
dhyāna-yoga, 114, 116
dialectic, Dualism is essence of, 18
dimensions of *manda* and *śīghra* epicycles, 289
dissipation of sleep inertia, 99
distinction assumption, 35
distinctive phonetic quality of Sanskrit, 276
distortion of
  objectivity issue, 30
  subjectivity issue, 30
dominant mode of cognition, 18
door to mystery, 37
dorsal occipito–parietal pathway, 94
double-blind technique, 13
dream state, 192
dual signalling schemes, 49
dualism
  is an excellent tool, 22
  is essence of dialectic, 18
  process of, 12
dualistic
  America, 12
  assumption of separateness, 14
  conceptions, continual correction of, 33
  models of common belief, 47
  science, 13, 34-35, 38
  tendency(ies), 24, 28
duality of objectivity vs. subjectivity, 29
dynamic range for quantification of consciousness, 254

Eastern
  culture, 2
  perspectives, 72
  philosophical traditions, 72
  religions, 246
  traditions, 70
Eastern world,
  culture of, 2
  philosophy of, 2
eccentric and the epicyclic models of planetary motion, 291
eclipse, 298
ecological perspective, 98
Eddington, Sir Arthur, 141
effective computation, model of, 47
effectiveness of alternative medicine, 166
Einstein, Albert, 6, 132, 140-42, 159, 161, 245, 250, 252
  Relativity Theory of, 131, 139
ekādisaṁkhyāviśeṣa, 308
Ekman, Paul, 234-35
Electrodermal recordings, 63
electro-encephalography (EEG), 72, 118-19
  rhythms of brain, 224

electromagnetic waves, 184
eleven-dimensional space, 157
emotion, expressions of, 234
emotional
  control, 80
  information, 80
  insight, 80
  intelligence, 80, 122
energy-information, 113
engine of non-dual science, 36
Engstrom, David, 111
enlightened
  jīva, 148
  matrix of *turīya*, 72
epistemological problems, 5
equinoxes,
  motion of, 289
  position of, 299
Erikson, Erik H., 106, 121
ersatz scepticism/scientism, 26
essence and via regia of science, 15
essential
  intrinsic consciousness, 113
  qualities of great scientist, 25
ethnocentrism, 20
Euclidean space, 146
European Renaissance, 3, 5, 243
evolution of
  adaptive conscious behaviour, 109
  humanity, future of, 226
evolutionary state of consciousness, 230
executive control of behaviour and mentation, 89
existence,
  concept of, 231
  holistic process of, 125
  of Reality, primordial state of, 240
existential
  ecosystemic mode, 108
  living, 104
experience of no experience, 123
experienced
  O-E-T, 137
  time, 149
experiences,
  assimilation of, 101
  field of, 148-49
experimental
  analysis of behaviour, 28
  testing of hypothesis, 21
experiments of animal intelligence, 52
expressions of emotion, 234
extrasensory perception, 60

features of tesseract, 200
field of
  acoustics impacts, 279
  experiences, 148-49
first-order collection of neural patterns, 102
five elements of universe, 277
Flamm, Bruce, 31
fly-on-the wall, 164
forces of nature, 210
Foundation for Research on the Nature of Man, 16
four
  stages for production of sound, 265
  states of human consciousness, 75

four-dimensional
  arrangement of universe, 195
  continuum, 175, 186, 193, 198-99, 201, 203-04, 206, 208-09, 216, 218-19, 221, 229
  mathematical space, 176, 178
  nature, 203
  space–time, 159
    of physics, 157
  time–space continuum, 145-46
  universe, 193, 195
fourfold
  levels of speech, 266-67
  Vedic model for sound field, 279
fourth state of consciousness, 74
fraction-existing, nature of, 221
fraction–living continuum, 212
  concept of, 187
frame of awareness, 236
framework of modern neuro-science, 71
fronto-temporal dementia, 103
functional magnetic resonance imaging (fMRI), 72
  techniques, 161, 166
fundamental
  assumptions of Vedic concepts, 10
  emotional responses, 80
  questions of consciousness, 232
Fundamental Law of Spirituality, 147, 165
future of evolution of humanity, 226

Gamma activity, 119
Gandharva Veda, 260
Gardner, Martin, 176, 178
Gauḍapāda, 75-76
Gāyatrī *mantra*, 272
general
  arousal, 81
  concept of existence, 231
  feature of neural networks, 56
genetic endowment, storehouse of, 85
geographical reasons, 2
geometrical picture of planetary motion, 284, 289, 300
Gheraṇḍa Saṁhitā, 115-16
Gibson, J.J., 57
  ideas of, 57
gist consciousness, 123
Gītā, 144, 161
global
  matrix function, 93
  warming, 18
God Almighty, 245
Goddess of Speech (Vāgdevī), 258, 272
Goddess Sarasvatī, 270-72
Goetsche, P.C., 37
Golasāra, 284, 286
goodness
  and knowledge, 147
  of self, 146
  knowledge of, 146, 148
Goswami, Amit, 131
grace of Dāmodara, 285
grammar, philosophers of, 265
grasp of rational mind, 241
great luminaries of twentieth-century science, 244
great scientist, essential qualities of, 25
Greek philosophies, 3, 243

Greene, Brian, 140, 145
Griffin, Donald, 28
Griffiths, Bede, 33
gross–subtle–causal body complex, 153

Hanumān, 121
Hawking, Stephen, 8, 247
heart of
  non-dual science, 34
  scientific methodology, 18
Hegel, 18
Heisenberg, Werner, 1
herbal chemistry, 10
Herrnstein, R.J., 50
Herson, Michael, 28
higher level of consciousness, 252
higher will, 133
higher-dimensional spaces, 177
Himalayan range of mountains, 2
Hindu
  definition of reality, 252
  description of total consciousness, 252
  philosophy, 251-52
  view of time and consciousness, 251
  worship practices, 275
Hinduism, 68, 70, 246, 252
  classical literature of, 257
hiraṇyagarbha, 158
history of
  astronomy, 283
  science, 292
holistic
  process of existence, 125
  superconsciousness, 124
homeokinetic(s), 108, 110
  activity, 113
  system, 111
homoeostasis, 108-10
  mechanism (HSM), 222
Honda, M., 264
Hrobjartsson, A., 37
human
  being, lifetime experience of, 253
  civilizations, advancement of, 173
  memory, storage of, 182
  neocortex, 58
  state of consciousness, three levels of, 235
human consciousness,
  concept of, 80
  four states of, 75
  known states of, 75
human sensual faculties, 8
  limitation of, 247
humanity, ignorance of, 212
hypothesis, experimental testing of, 21

Iberall, A.S., 110
idea(s) of
  knowing truth, 244
  Gibson, 57
ignorance of humanity, 212
illusion of three-dimensional continuously moving universe, 184
I-me-mine, sense of, 84
inaccurate misrepresentation of the complexity of subjectivity, 30
incremental experienced time, 145

Indian
astronomy, 298
cultural traditions, 2
philosophical systems, 3, 243
philosophical tradition, 293
planetary theory, 283
thoughts, 243-44
indulging mode, 137
infinite dimensions of consciousness, 248
influence of
consciousness, 66
laws of nature, 210
Vedic thoughts, 5
information, objectivity of, 1
informational relationship, 107
inner
environment, 72
planets, motion of, 291
insentient O-E-T, 158
instruments of modern science, 2
Integrated Information, 106-07
Theory, 107
integrated self-in-reality, 71
integrative mode of blissful conscious, 68
integrative process of living organism, 106
intellectual states of *jīva*, 137
intelligence, beam of, 154
interactive wakefulness, 71
internal faculty of perception, 85
intricacies of ancient knowledge, 5
intrinsic
awareness, 83, 123
consciousness, 78
inputs, origin of, 232
introspectionism, 36
introspective consciousness, 95
intuitions of sages, 66
intuitive consciousness, 97
inverse relationship between love and suffering, 147
ipseity, 94-96
Islamic or the Greco-European traditions of astronomy, 283
Īśopaniṣad, 241
Īśvara, 133, 135, 138, 153, 162
timelessness of, 139
will of, 135
Iversen, J., 119

Jata, Ravi, 189, 192, 198, 200, 208, 213, 221
Jha, G., 10
jhāna, 115
jijñāsā, 300, 307
Jitatmananda (Swami), 131
jīva(s), 78, 130-33, 135, 138, 144, 146, 148-49, 151, 154, 162-66
intellectual states of, 137
perspective of, 152
prerogative of, 162
reflected consciousness of, 154
spiritual nature of, 137
jīva-bhāva, 121
jñānendriya(s), 83-84
Jñāneśvara, 69
Judeo-Christian religio-spiritual tradition, 18
Jyotirmīmāṁsā, 286, 293
Jyotiṣaśāstra/*Jyotiḥśāstra*, 285-86 293

Kabīr, 240
Kahneman, Daniel, 23
Kaivalya, 88
kakṣyāmaṇḍala, 290
Kali era, 297
kalpa, 289, 291
Kaṇāda Ṛṣi, 277
karma-yoga, 144
karmendriya(s), 83-84
karṇa(s), 303-05, 307, 314
karuṇa, 112, 269
karuṇā, 68, 70-71
kaṭapayādi notation, 296
Kaṭha Upaniṣad/Kaṭhopaniṣad, 78, 202, 240
Katz, Jerry, 32
Kauṣītaki Upaniṣad, 78
Keizer, G., 280
Kelso, Scott, 111
Kenopaniṣad, 204
Kerala mathematicians, 302
Kerala school of Indian astronomy and mathematics, 282
Kinetic theory of gases, 253
Kirkpatrick, R., 3, 243
knowing truth, idea of, 244
knowledge
  of goodness, 146, 148
  of self, 144, 146, 148
  passive transmission of, 295
known states of human consciousness, 75
Koch, C., 235-36
koṭi, 303, 307, 309-14
Krishnamurti, J., 69, 114, 129, 241
Kṛṣṇa, Lord, 270-72
kṣetracchedana, 301
kṣipta, 87
kulturkries, 5
Kwang Cha, 31

lagna, 298
Lamm, C., 166
land of winners vs. losers, 20
language of science and scientific approach, 9
language, acoustical aspects of, 276
language-functioning process, 61
latitudinal motion, period of, 283
Laureys, S., 113
Laws of Form, 35
laws of nature, influence of, 210
Leibniz, G.W., 47
Libet, Dr., 161
life satisfaction scale, 166
lifetime experience of human being, 253
limitation of human sensual faculties, 247
Lindsay, R.B., 259
Linguistic research on chimpanzees and bonobos, 45
literature, acoustical aspects of, 276
living organism, integrative process of, 106
Llinas, R.R., 103
Lobo, R.A., 31
logical
  computation, 46
  derivations, 254
  fallacy, 161

way of living, 174
Lutz, A., 119, 236

𝒨-space, 138-39, 142-43, 146, 149, 154, 163
Mādhava, 282
madhyamā, 264-67, 279
level, 268-69
Magnetic Resonance Imaging (MRI), 118
magnetoencephalography (MEG), 72, 118-19
magnificent mystery, 103
Mahābhāskarīya, 298
mahat, 138-39, 142, 146, 148, 154, 158
Maitrī Upaniṣad, 115
major
concepts of modern physics, 197
conscious mental functions, 88
cultural and philosophical differences, 243
scriptures of *dharma*, 193
management of
neurobehavioural energy, 122
manaḥ, 84-85, 113
manas, 72, 87
manascakra, 87
manas-cakṣu, 85
manda and *śīghra* epicycles, dimensions of, 289
manifestation of *Brahman*, 242
mantra(s), 258, 279
collection of, 257
melodious pronunciation of, 260
proper recitation of, 261
recitation of, 268
Māṇḍūkya Upaniṣad, 70, 75
Marx, Karl, 18
material realm, 147
mathematical
derivations, 254
discovery, joy of, 302
model of spirituality, 151
matrix
awareness, 80
consciousness, 113
māyā, 131, 134, 138, 153, 214, 241, 252
concept of, 229
measuring rods of spiritual progress, 138
mechanistic model of computing, 47
meditation, 86
practices of, 70
meditative
brain, 118
traditions, 65
melodious pronunciation of *mantras*, 260
meta-consciousness, 122
metaphysical truths, 4, 243
mind,
basic ingredients of, 107
conceptualization of, 48
scientific study of, 88
mindfulness meditation, Buddhist way of, 69
mind-intellect, 134
Minkowski Metric, 142, 145
Mirsky, S., 252
Mishra, D., 10
mistaken identification of spirituality, 167
Mitka, M., 31

modality-specific
  information, process of, 82
  stimuli, 97
  thalamo-cortical network, 82
model of
  consciousness, 248
  effective computation, 47
  time, 250
modern
  acoustical view, 268
  chemistry, 3
  neuroscience, framework of, 71
  neurosciences, 130
  physics, relativity theory of, 130
  quantum theory, 35
  scanning techniques, 174
  scholar of Upaniṣads, 75
  view of speech production, 264
modern science,
  conclusion of, 135
  definition of, 244
  instruments of, 2
modern scientific
  conception of cosmos, 141
  languages, 247
  methods, 3
modes of being-in-existence, 107
modulation of network, 57
motion of
  equinoxes, 289
    description of, 298
  inner planets, 291
  planet, centre of, 291
Mudaliar, A. Devaraja, 162
Mukhopadhyaya, S., 265
mukta, 112-13
mukti, 88
mūlādhāra cakra, 265
multicellular, 109
multimodal associative cortex, 88
Muṇḍakopaniṣad/Muṇḍaka Upaniṣad, 137, 200
music, acoustical aspects of, 269
Myers, Ronald, 63
mystery,
  depths of, 38
  key, of, 32, 34
mystical, sensation of, 245

Nāda-Brahman, 262-63, 277
narcissistic tendencies, 166
narrowed interpretation of truth, 25
National Geographic, 27
National Institute of Health, 166
National Institutes of Health, National Centre for Complementary and Alternative Medicine (NIH-NCCAM), 26
nature of
  consciousness, 153
    host, 209
  continuum, 216
  fraction-existing, 221
  science, 291
nature, forces of, 210
nature's intelligence, 65
neural
  building blocks, 58
  computing, 57
  learning, 54

networks, 57
  general feature of, 56
  patterns, first-order collection of, 102
neurobehavioural energy, 124
neurobiological
  and cognitive domains, 103
  processes, 80
neuro-ecology, 108
neurosciences, 80
new model of
  consciousness, 94
  self-consciousness, 94
new way of living, 174
Newberg, A.B., 119, 236
Newtonian
  definition of scientific truth, 1
  or "clock work" universe, 132
Noe, Alva, 156
non-dual
  scientist, 35, 37-38
  vision of connectedness, 38
    relatedness, 38
non-dual science, 14, 34-39
  engine of, 36
  heart of, 34
non-dualism, 33
non-dualistic philosophy, 75
non-duality, 33
  is harmonious, 34
  proposed definitions of, 32
non-human cognition, 48
Non-Rapid Eye Movement Wakefulness (NREM-W), 71
non-REM
  alertness, 124
  sleep, 124
normalization of consciousness, 208
Northrop, F.S.C., 1
objective
  causality, world of, 45
  knowledge, 44
objectivity
  issue, distortion of, 30
  of information, 1
  of observation, 29
  vs. subjectivity, duality of, 29
objects of consciousness, 234
objects–emotions–thoughts, 132, 152, 154
observation, objectivity of, 29
Occam's Razor, 36
ocean of unbroken Bliss, 134
O-E-T, 134-35
old paradigms of
  autonomy, 47
  determinism, 47
omnipresence of Supreme or *Brahman*, 241
one is light to oneself, 83
oneness of continuum, 193
optimal arousal, 92
oral transmission of Veda *mantra*s, tradition of, 261
Oriental Manuscript Library, Kerala University, 296
origin of
  intrinsic inputs, 232
  various philosophies, 115
  various religions, 115
  Yoga, 115
original matrix of Vedic Hinduism, 69
overarching role of Consciousness in Vedānta, 131

Padoux, A., 279
parā, 264-67, 269, 279

Para-Brahman/Param Brahman, 199, 262
Paramātmā/paramātman, 83, 117, 175, 194, 199, 201-05, 208, 229
  definition of, 200
  realization of, 229
Parameśvara, 283, 298, 300
parapsychology, 16
particle nature of continuum, 187
passive transmission of knowledge, 295
paśyantī, 264-67, 279
  level, 268-69
Pātañjala Yoga-Sūtras, 10, 115
Patañjali, 69, 77, 86-87, 117
Patel, A.D., 264
Penrose, Roger, 8, 155, 247
perception of Realized Being, 134
perception, internal faculty of, 85
Perfect Being, 147
period of latitudinal motion, 283
personal time, 149
perspective of *jīvas*, 152
pet theories, 37
phenomenal
  consciousness, 95
  experience, 80
  world, 157
phenomenon of consciousness, 42
philosophers of grammar, 265
philosophical
  assumptions, 2
  influence, 2
  traditions, 2
philosophy of
  Eastern world, 2
  Western world, 2
philosophy, acoustics in, 277
physical
  space–time, 158
  time, 149
  world, 157
physics and Vedānta, synergy between, 131
picture of planetary motion, 298
Planck's time, 181, 184
planetary
  computational scheme, 298
  longitudes, computation of, 287
  motion,
    geometrical picture of, 284, 289, 300
    picture of, 298
  positions, computation of, 288, 298
Popular Mechanics, 15
Popular Science, 15
position of equinoxes, 299
Positron Emission Tomography (PET), 118
Prabhupada, Bhaktivedanta Swami, 207
practice(s) of
  dhyāna, 114
  meditation, 70
  Yoga, 77, 86
prajñā, 68, 70, 75-76, 88, 112
prakṛti, 82, 160
  forces of, 135
pramāṇas, 293
prāṇa, 124, 238

Prapañcasāra Tantra, 266
Prasad, M.G., 279
Pratibimbavāda, 154
pratimaṇḍala, 290
pratyakṣa, 293
pre- and post-central gyri, 119
pre-conscious processing, 124
pre-consciousness, 124
prediction, centralization of, 103
prefrontal–cingulate–parietal cortex, 94
prefrontal–parieto–temporo–occipital neural network, 124
pre-requisite for consciousness, 44
prerogative of *jīva*, 162
presence, concept of, 96
primary source of synergy, 131
primordial state of existence of the Reality, 240
process of
  dualism, 12
  modality-specific information, 82
proper recitation of *mantra*s, 261
proposed definitions of non-duality, 32
prosopagnosic
  patients, 62
  responds, 63
proto-self, 102
psychological
  process of speech formation, 267
  refractory period, 98
psychometric instruments, 166
pure
  consciousness, 238
  existence, 238
purity of scientific research, 32
purpose of consciousness normalization, 212
pursuer of science, 1
pursuing truth, Vedic ways of, 4
pursuit of scientific enquiry and experiments, 4
puruṣa, 77, 82, 85, 193-94, 196, 199, 202
  arrangement, 197
puruṣa–prakṛti, total reality of, 83
puruṣārtha, 162
Puruṣa-Sūkta, 164, 175, 192, 199, 201
Pythagorean
  scale, 272
  Theorem, dissectional proof of, 301
  triplets, 295

qualia, 107
quantitative domain of science and technology, 235
quantum
  characteristics, 46
  mechanical approach, 54
  mechanics, 48, 65, 131
    tangled computations of, 48
  models, striking success of, 55
  neuroscience, 108
  relativity, 131
  theoretic models of brain function, 54
  theory, 55
question of

adaptive organization, 57
consciousness, 42, 46

Racism, 20
Radhakrishnan, S., 75
radical dualism, 17, 19, 24, 26, 28, 30, 32, 37, 39
of America, 16
radically dualistic, 18
environment, 13, 23
Raichle, M.E., 96, 167
Rāma (Lord), 121
Ramabhadracharya, N.S., 264, 266
Rāmadāsa, 69
Ramakrishna Paramahamsa, 69, 237
Raman Maharshi, 69
Raman, Sir C.V., 258
Ramasubramanian, K., 282, 284, 297
Rao, Suvarnalata, 270
rasa, different connotations of, 270
rational mind, grasp of, 241
Ravi, 285
reality, Hindu definition of, 252
realization,
of Being, 74
of *Brahman*, 73
of *Paramātmā*, 229
spiritual fulfilment of, 262
time of, 163
Realized Being, perception of, 134
recitation of *mantra*s, 268
reductionist theory of view, 7
re-establishment of basic wakeful awareness, 99
reflected consciousness of *jīva*, 154
Reflected Consciousness, Theory of , 154
regulation of complex neural processes, 56
Relative Awareness, 147
relative motion in space–time continuum, 150
relativity
physics, 140, 147, 149
theory of modern physics, 130
REM sleep, 237
Republican(s), 13, 19-20
reunion of consciousness, 226
revolution in scientific thought, 130
Ṛgveda, 42, 74-75, 175, 238, 242, 259, 261
Rhine, J.B., 16
Roman Empire, 243
Rosenthal, Robert, 30-31
experimenter effects of, 13
Rubin, D.B., 30

Sacks, Oliver, 60
Sadguru, Yogī Śrīraṅga, 262, 269
sages, intuitions of, 66
sahasraśīrṣa, 193
Sāma, 74, 261
samādhi, 87, 114, 117
successive stages of, 88
sameness, abstract concept of, 53
saṁkhyā, 301
Sāṁkhya system, 82
samprajñāta, 88
Sanātana Dharma, 175, 177,

191, 199, 229-30, 257
Saṅgamagrāma, 282
Saṅgīta Ratnākara, 262, 265
Śaṅkarācārya/Śaṅkara, Ādi, 69, 75-76, 93, 154, 237
Sanskrit,
acoustical aspects of, 276
distinctive phonetic quality of, 276
language, 276
Sarma, K.V., 282, 286, 296-97
Śārṅgadeva, 262, 265
sat-cit-ānanda, 68, 70, 73, 121
Satya, Y.S., 7, 246
scholars of Vedānta, 245
schools of Indian thoughts, 4
Schrödinger, Erwin, 5, 7, 47, 55, 244, 246
science,
and scientific approach, language of, 9
and spirituality, conflicts between, 227
and technology, 173-74
quantitative domain of, 235
history of, 292
nature of, 291
scientific
analysis of consciousness, 231
attitudes toward consciousness, 42
enquiry and experiments, pursuit of, 4
enterprise, 15
methodologies, 10
methodology, heart of, 18
paradigm of age, 48
pursuit, 1
research, purity of, 32
scepticism, 25
study of
brain, 88
mind, 88
truth, 4
Scientific American, 252
Seal, B., 277, 279
Secret Service agents, 235
self,
conscious awareness of, 111
knowledge of, 144, 146, 148
self-actualization, 105
self-as-object, 101
self-as-subject, 100
self-consciousness, new model of, 94
self-integrity, wisdom of, 105
self-qualia, 95
self-realization of Truth, 88
semantic-conceptual field, 89
sensation of mystical, 245
sense of
blissful self-existence, 73
ceaseless wonderment, 68
I-me-mine, 84
sensory and psychological stimuli, 166
sentient *jīvas* of cosmos, 136
separateness,
assumption of, 38
dualistic assumption of, 14
seven yogic *cakras*, 267
Sharma, D.R., 239-42, 263-65
Shringy, R.K., 263, 265
Siddhānta-Darpaṇa, 283-87, 291, 293, 296-97, 300-01
similarities between Vedāntic and relativity physics models, 142
singular nature of continuum, 229

Śiva Saṁhitā, 115-16
Sivananda (Swami), 279
Skinner, B.F., 28
  boxes, 50
sleep inertia, dissipation of, 99
Smith, Huston, 25
Smith, Vernon L., 23
Smythies, John, 156-59
Snyder, A.Z., 96, 167
social psychologist, 24
Society for Neuroscience, 68
Socrates, 18
Somayājī, Nīlakaṇṭha, 283-84, 286, 290-302, 314
  compositions of, 287
  paramguru of, 298
Soodak, H., 110
Sound Field, 261
source of bliss, 262
space and time, agency of, 238
space–time continuum, 142, 148-49
spatial information, 103
spatiotemporal patterns of activation, 56
Special Relativity Theory, 141
spectrum of *vāsanā*s, 138
speech
  formation, psychological process of, 267
  production, modern view of, 264
  fourfold levels of, 266-67
  Vedic view of, 268
Speeth, K.R., 122
Sperry, Roger, 63-64
spiritual detachment, degree of, 138
spiritual
  education, 167
  fulfilment of realization, 262
  level of *ātman*, 269
  literature, 279
  nature of *jīva*, 137
  progress, measuring rods of, 138
  realm, 147
  truth, 165
spirituality,
  acoustics in, 277
  mathematical model of, 151
  mistaken identification of, 167
split-brain patient, 64
spontaneous
  consciousness, 114
  electromagnetic activity, 119
  essence, 125
  neurobiological phenomenon, 95
Squire, L.R., 233
Sri Aurobindo, 69
Śrī Ramaṇa, Maharṣi, 152, 161
Srinivas, M.D., 282-84, 297
Sriram, M.S., 283-84, 297
standard bio–psycho–social model of mind, 104
Stannard, Russell, 155
Stapp, H.P., 8, 247
state of
  complete consciousness, 232
  continuum, 198
Sternberg, R.J., 121
sthita-prajña, 74, 88, 108, 112, 121
storage of human memory, 182
storehouse of genetic endowment, 85
striking success of quantum

models, 55
string theorists, 157
study of consciousness, 54
subconscious mind, concept of, 86
subfields of science, compartmentalization of, 13
subjective experience of absorption, 119
subjectivity issue, distortion of, 30
subject–object duality, 93, 107, 241-42
subtle differences in anatomy, 58
successive stages of *samādhi*, 88
śūnyatā, 123, 240
superconscious, 70
  equanimity, 121
superconsciousness, 72, 123
supernatural concepts, 174
supervised learning, 57
Supreme Being, 75, 277
sustaining philosophies, 2
suṣumnā, 266
suṣupti, 75
svajanmakāla-jñāpanārtham, 299
svapna, 75
svaras, 269
svārtha, 83
svayam jyotiḥ, 95, 112
Swahananda (Swami), 154
Swanson, L.W., 232
synergy
  between physics and Vedānta, 131
  primary source of, 131

Tagore, Rabindranath, 69
Taittirīya Upaniṣad, 261
Talmudic sources, 18
tamas, 83
tāmasic *vāsanā*s, 162
tangled computations of quantum mechanics, 48
Tantra, 272
Tantrasaṁgraha, 286
Taoism, 68, 70
Tchebychev's Inequality, 23
Tejomayananda (Swami), 138
temporal organization of behaviour, 103
tendencies in Vedānta, 134
Teresi, Jean, 166
tesseract, 178-80, 199, 229
  features of, 200
thalamo–cortico–striate
  circuits, 93
  system, 88
theoretical basis for using time and temperature, 231
theory of consciousness, 43, 45, 66
Theory of Reflected Consciousness, 154
thermodynamic cyclic conception of life, 110
three
  essential steps of Yoga, 117
  levels of human state of consciousness, 235
  states of consciousness, 191
three-dimensional
  continuously moving universe, illusion of, 184
  host, 218
  objects, 226
  space, 177, 180, 186, 201, 203

spatial universe, 185
static behaviour, 187
universe, 180, 186, 195, 201, 208
representation, 198
world, 179, 196, 204, 206, 208, 213
Tibetan
Buddhists, 236
monk, 235
time,
and consciousness, Hindu view of, 251
model of, 250
of realization, 163
Vedāntic view of, 151
timelessness of *Īśvara*, 139
Time, Space, and Consciousness, 157
time–space continuum, 140, 143, 145-46
Todeschi, Kevin, 139
Tononi, G., 107, 113
total consciousness, 251
Hindu description of, 252
total reality of *puruṣa–prakṛti*, 83
tradition of
dharma, 9
oral transmission of Veda *mantras*, 261
traditional planetary theory, 283
transpersonal awareness, 70
true nature of universe, 175
true scientific scepticism, 26
truth,
narrowed interpretation of, 25
self-realization of, 88
turīya, 70-71, 74-76, 121, 123-24, 230, 238
unique conscious experience of, 76
twentieth-century science, great luminaries of, 244
two abstract poles of unique structure, 96
two-dimensional spaces, 136
two-party political system, 18
Tyagisananda, Swami, 176, 199

ubhaya sākṣī, 87, 122
becomes *ātma-tuṣṭa*, 124
ultimate existential fears, 105
Ultimate Reality (*Brahman*), 131, 239, 241
ultimate universal reality, 94
Umesh, G., 7, 246
unbroken Bliss, ocean of, 134
Uncle Sam, concept of, 104
Underwood, Lynne 166
undivided wholeness, Vedāntic idea of, 241
unique
blissful conscious being, 74
conscious experience of *turīya*, 76
structure, two abstract poles of, 96
universal field of consciousness, 66
universe,
five elements of, 277
four-dimensional arrangement of, 195
true nature of, 175
unusual sensory discrimination, 60
Upaniṣadic knowledge, 3, 243
Upaniṣads, 6, 42, 69, 147, 159,

200, 241, 270, 277-78
ūrdhva, 196

Vāgdevī, 272
vaikharī, 264-67, 269
  level, 268
Vaiśeṣika philosophy, 2
Vaiśeṣika-Sūtras, 277
Vaiśvānara, 75
vāk, 113
validity of application of *trairāśika*, 299
various
  philosophies, origin of, 115
  religions, origin of, 115
*vāsanā*s, 134-35, 138, 144, 153
  spectrum of, 138
Vaṭasseri, Parameśvara, 282
Vatican, 5
Vedānta, 69-70, 72, 77, 79, 83, 86, 121, 138, 140, 147, 151, 153, 156, 160, 163, 165, 239-40, 242
  basic teachings of, 137
  scholars of, 245
Vedānta philosophy, 82
  of life, goal of, 85
Vedāntic
  idea, 139
    of undivided wholeness, 241
  Indian literature, 108
  literature, 86
  model, 142, 158-59
  philosophy of life, 94
  teachings, 130, 158
  view of time, 151
Vedāntic concept(s), 131-32
  of cosmic mind, 134
  of spontaneous sound of peaceful silence, 113
Vedāntins, 133
Vedic
  chanting, 261
  communities, 261
  concepts, fundamental assumptions of, 10
  culture, 5
  fourfold
    level speech production, 269
    model for generation of speech, 257
    model of speech, 268
  Hinduism, original matrix of, 69
  ideas, 5
    of zero and infinity, 254
  knowledge, 2-4, 9, 243-44
    contribution of, 8, 247
  literature, 94, 120, 257, 262, 264, 276
  *mantra*s, 269
  philosophy, 10
  practices, 3
  psychology, 120
  quotations, 126
  scholars, 94
  seers, 69, 105
  texts, 42
  thoughts, 244
    influence of, 5
  traditions, 8-10
  view, 42-43, 264
    of production of speech, 268
    of speech, 268
  ways of pursuing truth, 4
ventral occipito–temporal pathway, 94
verbal information, 103
vidyārthī, 276
vikṣepa, 294

vikṣipta, 87
vīṇā, 270, 272
vipaśyanā, 68, 70
vīra, 269
virām-pratyaya, 88
virāṭ, 158
virtual
    continuum, 187
    host, 210
    nature of continuum, 208, 220
virtue of wisdom, 106
viśrānti, 295
visual agnostic patient, 62
Vivekacūḍāmaṇi, 237
Vivekananda (Swami), 69, 237

Wallace, B.A., 122
Wasserman, E.A., 51-52
Watson, I.K., 277
wave nature, 187
Wernicke's area, 89
Western
    culture, 2
    meditator, 235
    perspectives, 72
    philosophers, 132
    philosophy, 3, 243
    science, 17
    scientists, 5
    thoughts, 4, 243
Western world, 5
    culture of, 2
    philosophy of, 2
Westernized modern science, 4, 244
Weyl, Hermann, 155
What is Life, 246
Will of *Īśvara*, 160
Wirth, Daniel P., 31-32
wisdom
    of self-integrity, 105
    virtue of, 106
witness consciousness, 124
Woodroffe, Sir J., 279
working-memory representations, 103
world of
    ecologically valid interpretation, 23
    objective causality, 45
Wundt, Wilhelm, 35

Yajurveda, 74-75, 242, 258, 261
Yalom, I.D., 105
Yoga, 70, 115, 117, 121, 263, 272, 279
    basis of, 86, 115
    is skilful living, 94
    origin of, 115
    practice of, 77, 86
    three essential steps of, 117
yoga, 9
    practices of *dhyāna*, 10
Yoga Vāsiṣṭha, 138
Yoga-Sūtras, 69, 77, 86-87, 117
yuga(s)
    concept, 255
    cycle of, 251
yuktimūla, 295